Towards a Philosophy of Cosmic Life

David Bartosch · Attila Grandpierre · Bei Peng
Editors

Towards a Philosophy of Cosmic Life

New Discussions and Interdisciplinary Views

Springer

Editors
David Bartosch
Institute of Advanced Studies
in Humanities and Social Sciences
Research Institute for Globalization
and Cultural Development Strategies
Beijing Normal University at Zhuhai
Zhuhai, China

Attila Grandpierre
Budapest Centre for Long-term
Sustainability (BC4LS)
Budapest, Hungary

Bei Peng
Institute of Advanced Studies
in Humanities and Social Sciences
Research Institute for Globalization
and Cultural Development Strategies
Beijing Normal University at Zhuhai
Zhuhai, China

ISBN 978-981-99-2130-0 ISBN 978-981-99-2131-7 (eBook)
https://doi.org/10.1007/978-981-99-2131-7

This Springer imprint is published by the registered company Springer Nature Singapore Pte Ltd.
The registered company address is: 152 Beach Road, #21-01/04 Gateway East, Singapore 189721, Singapore

Contents

Editors and Contributors

About the Editors

David Bartosch (Germany) is an internationally recognized expert in Chinese, German, and other philosophical traditions. He specializes in transcultural comparative philosophy and has applied this knowledge to the fields of intercultural relations and ecological thought. He is a distinguished research fellow at Beijing Normal University at Zhuhai in the Guangdong–Hong Kong–Macao Greater Bay Area in China.

Attila Grandpierre (Hungary) is an internationally renowned solar astronomer, quantum physicist, biologist, cultural historian, musician, and poet. He was a senior research fellow at the Konkoly Observatory for many decades. At present, he is the director of the Hungarian Ecological Civilization Research Institute and the research president of the Budapest Centre for Long-Term Sustainability.

Bei Peng (China) is a musicologist, philosopher, and translator who is at home in both Eastern and Western cultures. After a long stay in Germany, she is now working in China at Beijing Normal University at Zhuhai in the Guangdong–Hong Kong–Macao Greater Bay Area. In her work, Peng uses music theory to develop a universal understanding of human existence on an aesthetic level.

Contributors

David Bartosch Institute of Advanced Studies in Humanities and Social Sciences, Research Institute for Globalization and Cultural Development Strategies, Beijing Normal University at Zhuhai, Zhuhai, China

John B. Cobb, Jr. Institute for Postmodern Development of China (IPDC), Claremont, CA, USA

Attila Grandpierre Budapest Centre for Long-Term Sustainability (BC4LS), Budapest, Hungary

Bei Peng Institute of Advanced Studies in Humanities and Social Sciences, Research Institute for Globalization and Cultural Development Strategies, Beijing Normal University at Zhuhai, Zhuhai, China

Chandra Wickramasinghe Buckingham Centre for Astrobiology, University of Buckingham, Buckingham, UK;
Centre for Astrobiology, University of Ruhuna, Matara, Sri Lanka;
National Institute of Fundamental Studies, Kandy, Sri Lanka;
Institute for the Study of Panspermia and Astroeconomics, Gifu, Japan

Chunyou Yan College of Education for the Future, Beijing Normal University, Zhuhai, China;
School of Philosophy, Beijing Normal University, Beijing, China

Introduction

David Bartosch, Attila Grandpierre, and Bei Peng

Abstract This introductory chapter to the book *Towards a Philosophy of Cosmic Life: New Discussions and Interdisciplinary Views* provides an overview on a multidisciplinary, multitheoretical, and transcultural dialogue among six authors from three continents. The aim of this discourse is to establish highly innovative perspectives and to explore new frontiers for the development of philosophical reflections and scientific foundations for the emergence of a common cosmic consciousness, an integral ecology, and a cooperative planetary civilization of humanity. The chapter includes brief profiles of the authors and chapter summaries. The golden thread running through the spectrum of ideas and points of view outlined here is provided by the underlying view that life is not an accidental emergence in an otherwise exanimate infinity, but an expression or manifestation of a self-unfolding cosmic life principle, namely as an implicit but central and general factor in the formation of everything.

What is life? Where does life come from? What is the cosmic principle of life? These are fundamental questions. They have continued to inspire human beings since ancient times. Whether in terms of mythology, religion, philosophy, science, psychology, or various other fields of study, and regardless their respective

All authors of this chapter are co-first authors.

D. Bartosch (✉) · B. Peng
Institute of Advanced Studies in Humanities and Social Sciences, Research Institute for Globalization and Cultural Development Strategies, Beijing Normal University at Zhuhai, Zhuhai 519087, China
e-mail: david.bartosch@bnu.edu.cn

B. Peng
e-mail: bei.peng@bnu.edu.cn

A. Grandpierre
Budapest Centre for Long-Term Sustainability (BC4LS), Budapest, Hungary
e-mail: attila.grandpierre@bc4ls.com

D. Bartosch et al. (eds.), *Towards a Philosophy of Cosmic Life*,
https://doi.org/10.1007/978-981-99-2131-7_1

linguistic and cultural environments—countless seekers have used their own words, ways of expression, and logic of thought to respond to these ultimate questions.

Our human physiology, especially our upright posture and bipedal gait, has enabled us to communicate face to face and also to turn our gaze to the sky, the original source of all the lights of our days and nights. Even the word 'under*standing*' is related to this. We are cosmic "offspring" *standing* in between the sky and the earth. Humanity forms a cosmic connection between the infinite "giver" of light, rain, and life—*Ouranos* (Οὐρανός) aka *Dyauspitar* (द्यौष्पितृ) aka *Tian* (天), etc.—and the planetary "womb" of Mother Earth—*Di* (地) aka *Prithvi* (पृथ्वी) aka *Gaia* (Γαῖα), etc.

Since ancient times, human understanding and the pathways of science have evolved as a collective tree of knowledge, bearing countless fruits and seeds based on the wisdom of many civilizations of many ages and places. Absorbed, reassembled, and enriched, for example, by ancient foundations and further discoveries made in Central and West Asia, many important philosophical and scientific achievements of East Asian, South Asian, and African civilizations were transplanted into the monastic cells and secret laboratories of medieval European thinkers. After the birth of the Renaissance, the discovery of the Americas, and due to further acquaintance with Asian philosophical lore, the Age of Enlightenment provided important impulses for many later philosophical and scientific developments all over the world.

However, the emerging dogmatic one-sidedness of the European mechanistic worldview of the 17th and 18th centuries, which identified everything as inanimate "material" and thus as a mere resource, or the (reductionist) positivism of the 19th and 20th centuries have fostered and accompanied a highly alarming process of creating artificial needs for the sake of disintegrating and, thus, meaningless power—leaving its destructive mark on history and the biosphere as a whole. The associated behavior patterns of non-sustainable economic exploitation disregard our cosmic interconnectedness and thus threaten the very existence of humanity today.

"But where danger lurks, the saving power also grows," the German poet Friedrich Hölderlin (1770–1843) once said. Since the 20th century, scientists have made the necessary progress to finally overcome the aforementioned misconceptions. The discoveries of quantum physics and related developments in biology and many other fields have inspired thinkers and researchers to seek more integrative, holistic, and open perspectives. These even provide certain points of contact with the wisdom of ancient peoples. It is now important and urgent that we transform our social systems according to such newly emerging *life-centered* worldviews. In this context, a new philosophical foundation must be developed. It has to be receptive to new perspectives in cosmology and biology, and it should have a transcultural foundation in humanity's many historical and modern traditions of thought.

So far, the necessary shift in our way of thinking has involved only a relative minority of observers. From an overall perspective, humanity is still influenced by pre-scientific worldviews in combination with non-sustainable forms of economy that implicitly correlate with outdated paradigms and views in the history of science. This makes it difficult to initiate the transformation that is needed to reach the common goal of the planetary family, namely, a long-term sustainable existence for all. Those in science who still resist holistic, non-dogmatic, and interdisciplinary thinking often simply adopt reductionist tendencies and project the image of the experimental system in their laboratory, which is a finite machine made up of parts, onto the superordinate reality of the infinite autopoietic nexus that we call the universe. General world views and models of life are still implicitly or explicitly dictated by the image of a man-made mechanism, which is a mere technological shadow of life, and by the processes of 'reification' (*Verdinglichung*) that have always privileged the few while crippling the lives of the many.

But there is hope for a change: space exploration has been in progress for the past 100 years, the field of SETI has evolved in parallel, and several thousand exoplanets have been discovered in the last 30 years. More recently, anomalies such as the interstellar object Oumuamua have even stimulated a new search for technosignatures of possible extraterrestrial intelligences, etc. The question of an animated universe beyond Earth is no longer limited to our neighboring planets or the moons of the gas giants of our solar system. Advanced scientific articles in astrobiology locate the origin of life in the interstellar reaches of space, while others theorize the possibility of non-carbon-based life forms. This presents us with an unorthodox and innovative question that may provide the philosophical impetus that we need to improve our situation and the situation of life on Earth in general. *Could life, after all, not only exist in the vast expanse of the universe, but be a central factor in shaping it?*

In our view, the open discussion of this fundamental question will be crucial to initiate a balance between human creativity and activity on the one hand and the living cosmic process that we call nature on the other. A new, cosmic understanding of life can provide the necessary traction to usher in a new era of life-centered global communities. The greatest danger to humanity lies in certain patterns of human social behavior itself. As long as we believe that human and non-human life is a random "mechanical" anomaly on a grain of sand in a boundless exanimate desert, the door to human caprice, selfishness, and self-destruction will be wide open.

In order to support the philosophical discussion, the editors and authors of this volume hope to provide the reader with more questions than answers and with more glimpses of the unknown than the known. We believe that such a philosophical approach is necessary to stimulate the intellectual process toward a future philosophy of cosmic life—concerning a sustainable and loving path for our species in the shared awareness of a living universe. Such a process of asking questions, spending a lifetime of seeking and finding answers, and courageously exploring the unknown is what makes human life truly meaningful. It is one of the ultimate tasks of our existence. In this sense, the term 'cosmic life' is used here as an open and comparative category in order to gather a plurality of different perspectives from different philosophical, scientific, thematic, historical, and cultural directions.

The authors of this volume are from five countries and have diverse multicultural and academic backgrounds in philosophy, astronomy, biology, musicology, and other fields. They contribute valuable insights that have been accumulated over many years in each case—*unique and forward-looking views on the relationship between human beings and the universe, life and the universe, and human beings and life.*

Our publication was motivated by an online meeting held on January 26, 2022, during which the six contributors presented talks on topics that are now the subjects of their chapters in this book. The global discussion was jointly initiated and organized by Dr. David Bartosch, Distinguished Research Fellow at Beijing Normal University at Zhuhai, Dr. Attila Grandpierre, Research President of the Budapest Centre for Long-Term Sustainability (BC4LS), and Dr. Bei Peng (彭蓓) of Beijing Normal University at Zhuhai.[1]

The first part of the meeting included an intense discussion presided over by the renowned US-American scholar and environmental activist Dr. John B. Cobb, Jr. (born 1925). It lasted more than three hours and resulted in a plethora of new philosophical perspectives on the nature of life and the cosmos. Cobb, who was born and raised in Japan, is the preeminent scholar of Whiteheadian process philosophy and the subsequent tradition of process theology, and one of the most important ecological and transdisciplinary thinkers of our time. His academic career spans more than 70 years and he has published more than 50 books. Cobb was a professor at the Claremont School of Theology and the Claremont Graduate University. He co-founded the prestigious Center for Process Studies, the Institute for Postmodern Development of China, and the Institute for Ecological Civilization. In addition, he was a visiting professor at numerous prestigious universities in the United States, Germany, and Japan. Cobb is the recipient of six honorary doctorates and is a member of the American Academy of Arts and Sciences.

The editors are very honored to present John B. Cobb, Jr.'s "Cosmic Life" as the first chapter of this book. The author develops a concept of life as living events, that is, as living processes expressing the property of novelty. He argues against a highly problematic physicalist reduction of physics and life. Topics such as the ontological status of the past, decision, and potentiality are included in the discussion. The entire cosmos is described as a process of 'Life Itself,' which is to be reflected as self-enriching and self-evolving in all its parts. As an inner ideal, the (practical) force of Life Itself provides personal guidance in our daily and extraordinary tasks and aspirations.

The second chapter, "Acknowledging Our Inalienable Links with the Cosmos," is authored by the well-known Sri Lankan-born British mathematician, astronomer, and astrobiologist Prof. Dr. Nalin Chandra Wickramasinghe MBE (born 1939). He had chaired the second segment of the authors' aforementioned online discussion. Wickramasinghe's academic career spans more than 60 years. His mentor was the famous astronomer Sir Fred Hoyle, and the two scientists worked closely together for more than four decades. Wickramasinghe was a professor at University College Cardiff, the University of Cardiff, and he currently holds a position at the University of Buckingham. He has also been a visiting professor in Sri Lanka and Japan. Wickramasinghe has authored or co-authored more than 40 books and more than 350 peer-reviewed scientific articles, many of which were published in prestigious journals such as *Nature*. Early in his career, Wickramasinghe made the groundbreaking discovery that some dust in interstellar space is largely organic, and he has developed a modern theory of panspermia over the past 40 years. He holds several honorary doctorates and has received many prestigious awards.

In twelve steps, Wickramasinghe unfolds a scientific and philosophical train of thought that considers the implications of accepting the concept of cosmic life. He argues for the cosmic origin of life by using probability calculations and discusses data supporting the theory of panspermia. The distribution of habitable exoplanets and the possibility of an interstellar exchange of life forms are discussed. Relevant studies of the controversial Polonnaruwa meteorite are covered, along with the history of research on meteorites, micrometeorites, and the idea of cometary bacteria. With regard to panspermia, stratospheric sampling and the idea of diseases and pandemics from space are presented as important areas of existing and future research.

The third chapter is titled "The Cosmic Life Instinct Points the Way to a Healthy Ecological Civilization" and is written by the Hungarian polymath and artist Dr. Attila Grandpierre (born 1951). He is the research president of the Budapest Centre for Long-term Sustainability (BC4LS). From 1974 to 2013, Grandpierre was a senior researcher at the Konkoly Observatory of the Hungarian Academy of Sciences and has published important works on solar astronomy. He was a visiting professor in the fields of quantum physics and theoretical biology at Chapman University in the United States and has held leading positions in important Hungarian institutions in the area of ecological thinking for many years. Grandpierre has developed a comprehensive approach that combines the natural sciences and perspectives of the humanities. He also published in fields such as astrobiology and Eurasian cultural history. To date, Grandpierre has published 17 books and many dozens of important scientific articles and book chapters in various fields. An active singer since 1975, he founded two legendary music groups and was awarded the Parallel Culture Prize in 2012.

Grandpierre's comprehensive perspective builds on the work of the Hungarian theoretical biologist Ervin Bauer, who mathematically derived all the basic biological equations from a single fundamental principle, known as the Bauer principle: Living organisms actively and continuously mobilize their free energy content to maximize their distance from lethal thermodynamic equilibrium. Grandpierre deepened the explanatory scheme of modern physics, extended the principle of least action to include biological teleology as a free variable and made this general action principle compatible with Bauer's principle. He further extended the so-obtained principle to include biological motivations playing the same role in biology as forces in physics. The resulting universal life principle is the most important step forward in our understanding of the structure and even the nature of the living universe itself. It provides the theoretical basis for a life-centered worldview and a cosmic ecology that embraces all forms of life. In this way, it offers a firm basis for an all-comprehensive, uplifting conduct of life in our individual and communal life, for good government, and for a healthy civilization.

The fourth chapter provides a discussion on "Holographic Philosophy as a Philosophical Basis of Cosmic Life Theory." It is authored by Prof. Dr. Chunyou Yan (严春友, born 1959) of the School of Philosophy at Beijing Normal University in Beijing and the College of Education for the Future at Beijing Normal University at Zhuhai in the Guangdong–Hong Kong–Macao Greater Bay Area in China. Yan is a leading expert in modern Chinese natural philosophy. He was born in Shandong Province, the home region of Confucius. From the beginning of his academic career, Yan sought to answer the question of the unity of the universe in the context of new discoveries in biology and cosmology. This led to the incorporation of traditional Chinese views on the relationship between the human being and the universe and to the development of Yan's holographic theory of cosmic life in many articles and books. This highly innovative theory incorporates relevant philosophical perspectives influenced by traditional Chinese Buddhism, Daoism, and Western philosophies, and it provides an important foundation for an East–West dialogue on the topic of cosmic life.

The main proposition of Yan's theory of cosmic holography, which was later renamed 'cosmic holographic unity theory,' is that all parts contain all the information of the whole. Yan's chapter describes the emergence of holographic philosophy in modern Chinese discourse, provides an analysis of the basic principles of holographic philosophy, and it points out its provability. He presents the inevitable conclusion of holographic philosophy that the universe is a vast living entity and that every part of it contains life. From this perspective, all things are to be seen as aspects of 'Universal Life' at different stages of evolution, and they are all evolving toward higher states of life. In terms of the current discourse, Yan's theory, which dates back to the 1980s and 1990s, is very relevant again.

Dr. Bei Peng (彭蓓, born 1979) is a Chinese musicologist who works at the Institute of Advanced Studies in Humanities and Social Sciences, Research Institute for Globalization and Cultural Development Strategies at Beijing Normal University at Zhuhai. She contributed the fifth chapter, which is titled "The Human Body as the Singing Universe." Peng studied musicology and philosophy in Germany. She received her doctorate from the University of Heidelberg on a key topic in the field of pre-modern Chinese music theory. This historical field of mathematical study was important not only as the foundation of court music, but also for the imperial system of metrology on which all Chinese units of measurement were based. Furthermore, the Chinese saw imperial court music as the acoustic embodiment of celestial laws. In relation to this, Peng has also explored Johannes Kepler's idea of 'world harmonics' (*harmonice mundi*) from a comparative perspective. In addition to these research interests, she is the co-editor of a Chinese edition of the works of Theodor W. Adorno and develops transdisciplinary perspectives in music philosophy.

For millennia, the idea that there is a universal order that connects human beings and the universe has informed cultures around the world. This order was often expressed in geometric, harmonic, or musical terms. From Pythagoras to Kepler, universal scholars firmly believed that this order represented the primordial code of all things. Peng explores a new interdisciplinary perspective that combines the fields of traditional Chinese medicine (TCM), music theory, and modern astronomical insights inspired by Keplerian world harmonics. On the basis of mathematical analysis, it is shown that the energy meridians of the human body correspond to a musical order, which in turn is inextricably linked to certain relational aspects of the celestial bodies of our solar system. Peng's research represents completely new discoveries in relation to this universal connection. Therefore, it also opens up new levels of understanding on the subject of cosmic life.

The sixth chapter, "Poly-contextural Cornerstones for a Transcultural Philosophy of Cosmic Life," is authored by Dr. David Bartosch (born 1977). He is a distinguished research fellow at the Institute of Advanced Studies in Humanities and Social Sciences, Research Institute for Globalization and Cultural Development Strategies at Beijing Normal University at Zhuhai. Most recently, he has also been a visiting researcher at the Budapest Centre for Long-Term Sustainability (BC4LS). A native of Germany, Bartosch received his academic training in philosophy and musicology in his home country, where he also began to work in interdisciplinary contexts early on. An expert on both German and Chinese traditions of thought, he is known for his transcultural comparative work on the history of European and Chinese philosophy, including thinkers such as Wang Yangming (1472–1529) and Nicolaus de Cusa (1401–1464). His general philosophical interests include topics concerning ecology, anthropology, cosmology, the world history of philosophy, cultural history, intercultural learning, and technology.

In his contribution, Bartosch provides a systematic and historical overview of important transcultural foundations for a comprehensive philosophy of cosmic life. Starting from the origin of the philosophical term 'cosmic life,' Bartosch moves on to Gotthard Günther's concepts of poly-contexturality and trans-classical science. These concepts are extended and complemented by the views of other thinkers to generate a new systematic foundation for what Bartosch calls 'panenbiotism' ("all-in-life-doctrine"). Against this background, Bartosch provides transcultural foundations for a future philosophy of cosmic life. Family-resemblant positions from ancient Hellenic, Chinese, South Asian, West Asian, and more recent European contexts are presented and discussed in transversal perspectives and also in view of some highly innovative hypotheses of our time. This discussion is structured in terms of three basic themes: 'cosmic life' as an expressed principle of incessant self-(re)creation through metabolism and partial self-consumption, 'cosmic life' as an integral feature of a vast or infinite universe, and, last but not least, 'cosmic life' as an integral aspect of cosmic 'light' in a broader sense.

The editors would like to thank Dr. John B. Cobb, Jr., Prof. Dr. Chandra Wickramasinghe MBE, and Prof. Dr. Chunyou Yan for their contributions. We are especially grateful to Dr. Niels Peter Thomas, Leana Li, Herb Zhang, and all of their colleagues at Springer Nature who were involved in the production process of this book. Furthermore, we would also like to thank Dr. Alexis Smith of Hanover College, Hanover, IN (USA) for proofreading two chapters of the book. Finally, the editors are grateful to their respective institutions, the Institute of Advanced Studies in Humanities and Social Sciences and the Research Institute for Globalization and Cultural Development Strategies at Beijing Normal University at Zhuhai (BNU Zhuhai) in China as well as the Budapest Centre for Long-term Sustainability (BC4LS) in the beautiful capital of Hungary, for their support.

Notes

1. Hungarian and transliterated Chinese names of editors and authors of this volume are written in the order of given names followed by family names. In the lists of references, transliterated names of modern Chinese authors are listed in the same way as the names of modern non-Chinese authors (initials of given name followed by family name). The names of classical or pre-modern authors are spelled out in full in the lists of references (e.g., 'Nicolaus de Cusa'). This also applies to the transliterated names of pre-modern Chinese authors, which are displayed in the traditional name order of the family name followed by the given name (e.g., 'Wang Shouren, 王守仁').

David Bartosch (Germany) is an internationally recognized expert in Chinese, German, and other philosophical traditions. He specializes in transcultural comparative philosophy and has applied this knowledge to the fields of intercultural relations and ecological thought. He is a distinguished research fellow at Beijing Normal University at Zhuhai (BNU Zhuhai) in the Guangdong–Hong Kong–Macao Greater Bay Area in China.

Attila Grandpierre (Hungary) is an internationally renowned solar astronomer, quantum physicist, biologist, cultural historian, musician, and poet. He was a senior research fellow at the Konkoly Observatory for many decades. At present he is the director of the Hungarian Ecological Civilization Research Institute and the research president of the Budapest Centre for Long-Term Sustainability (BC4LS).

Bei Peng (China) is a musicologist, philosopher, and translator who is at home in both Eastern and Western cultures. After a long stay in Germany, she is now working in China at Beijing Normal University at Zhuhai (BNU Zhuhai) in the Guangdong–Hong Kong–Macao Greater Bay Area. In her work, Peng uses music theory to develop a universal understanding of human existence on an aesthetic level.

Cosmic Life

John B. Cobb, Jr.

Abstract The present contribution is a new text which has been written under the impression of more than seventy years of active dedication to the innovative development of process philosophy. In this context, the topic of cosmic life is of utmost importance. In view of this, new terrain is explored here. The first main segment presents the view of life as living events, that is, as living processes which express the feature of novelty. The author argues against the paradigmatic, and highly problematic, physicalist reduction of physics and life. In a second step, the whole cosmos is defined as Life Itself. One of the guiding questions is: Does the past exist? Furthermore, the factors of decision-making and of potentiality are discussed in relation to this. In a third step, the whole cosmos is described as a process of Life Itself, which is to be reflected as self-enriching and self-evolving in all its parts. In the form of an inner ideal, the (practical) force of Life Itself provides personal guidance in all of our daily or uncommon tasks and aspirations.

Keywords Life Itself · Living events · Cosmic struggle for life · Novelty · Potentiality

The title of this book implies several dimensions of meaning. We could be asking whether the universe is composed ultimately of living things. We could be asking whether the cosmos is itself alive. We could be asking whether the cosmos actively supports life.

Life as Living Organic Events that Express Novelty

To discuss this topic, we need first to reflect on what we mean by life. Not long ago, scientists thought the natural world was divided quite simply into things that are alive

---

John B. Cobb, Jr. (✉)
Institute for Postmodern Development of China (IPDC), Claremont, CA 91711, USA
e-mail: jcobb@ctr4process.org

D. Bartosch et al. (eds.), *Towards a Philosophy of Cosmic Life*,
https://doi.org/10.1007/978-981-99-2131-7_2

and things that are inanimate. Plants and animals are alive, whereas machines and stones are inanimate. Of course, living things died and, thereby, became inanimate. Scientists assumed that even living things were composed of parts that were inanimate. They could be analyzed into atoms, entities that could not be divided, and the entities that made up the world studied by science were thought to be inanimate matter. Obviously, these were sometimes formed into organisms.

Today, a strong case can be made that the atoms of which the natural world is composed all have an organic character. Whitehead described his thought as a philosophy of organism on this basis. Of course, the atomic organisms are very different from animals and plants. But they are even more different from lumps of matter. They are self-organizing momentary events. Whitehead calls them actual entities or, more informatively, actual occasions of experience. For him, to be actual is to be experiential. To be experiential is to have subjective being, to exist in and of itself. Inanimate matter, on the other hand, has no existence in and for itself. A stone is an example of 'inanimate matter.' As a stone, it exists only as an object for subjects. In itself its existence is only in the actual occasions of which it is composed.

The stone has mass, which is the technical meaning associated with what philosophically we call matter. Each of the occasions of which it is composed has mass. What possesses mass also possesses energy: $E = mc^2$. But physicists have increasingly discovered that much of the world they study does not have mass. If the basis for affirming matter as a constituent of reality is mass, then we must say that most of physical reality is not material. Most of the actual entities of which most of the visible world is composed is now understood to be composed of quanta which have no mass.

We used to think of much of space as a vacuum, but now we understand that something is happening in the vacuum. These events, constituting plasma, are just as real as material events. They are just as 'physical,' if that means they are objects of study in physics. Like 'light' and the whole quantum world, they have energy but no mass. It seems that the real atoms, that is, indivisible entities, are pure energy rather than the pure matter of classical physics. If we can free the philosophical term 'physical,' from the connotations of 'material,' we can say that energy is just as physical as mass.

Now we are discussing 'life.' If we define life in such a way that whatever is self-organizing or is subjective is 'alive,' then everything is alive or can be analyzed into elements that are alive. If we make this move, then we can certainly affirm that life is foundational to the cosmos. But the ordinary distinction of living and nonliving things is lost. There is some advantage in focusing, instead, on what Whitehead calls societies of things.

Often the term 'organic' is identified with 'living.' Whitehead, on the other hand, distinguishes between occasions in which self-organizing introduces novelty and those in which it does not. He also wants to focus on what he calls 'societies of occasions,' that is, what we usually think of as things, sticks and stones, spoons and radios, plants and animals. Then a thing is alive if the novelty in its members has a significant effect on what it does. Living is a matter of degree. Even if there are some flashes of novelty in some of the actual occasions of experience of which the stone is composed, we can still say that the stone is inanimate. Even if a squirrel's behavior is quite predictable in a general way, we can still say it is very much alive.

In other words, it turns out that pure matter does not exist. We get the idea from the way that energy events are organized, and how they appear in touch and sight. But what is there, in the world, independent of us is ultimately energy, occurring in events. These events simply cannot be understood when we are looking for tiny little objects like stones. Perhaps we can say that the units of which the world is composed embody 'pre-life.'

Sadly, a materialist bias is still widespread among physicists. Step by step they are acknowledging that their modes of explanation do not account for much of the world. Because they are organized in multiple guilds, most of them can ignore the worlds of plasma and quanta. They just dismiss them as queer. But this is not a sustainable solution to metaphysical questions or even scientific ones.

There are scientists who try to make sense of the world. One of the best was David Bohm. Bohm recognized that the problem lay in the Indo-European languages. The language in which we express ourselves shapes our understanding. The Indo-European languages give primacy to the noun or pronoun around which most sentences are built. Even when we do not know what our noun or pronoun refers to, we posit it. For example, as Zhihe Wang pointed out to me, when there is rain, we say 'It is raining.' But we cannot say what the 'it' refers to.

Bohm himself found help in Buddhism. But he thought our best chance in the English language world was to get in the habit of starting sentences with a gerund. In this case we would say 'Raining is occurring.' The gerund refers to an event. In the quantum world, events are constantly occurring. One event flows into other events.

For anyone who wants to go deeply into an understanding of a world of events and processes out of which objects emerged, I recommend the philosophy of Whitehead. Whitehead worked out his thinking before much was known about the quanta. He dealt intensively with relativity, thinking that Einstein's mathematically successful formula did not make philosophical sense. Relativity points to relationships, and Whitehead's philosophy is all about relations.

Chinese should have an advantage in developing a science that is not distorted by the substantialist preference of Indo-European languages. In antiquity, the Chinese were able to appreciate and absorb the insights of Gautama. To this day Gautama's rejection of substances has not been understood in India.

Let me confess that I am myself guilty of following the guidance of the English language. I am writing an essay on 'life' even though I know that there is no such thing. There are events that have characteristics that justify the adjective 'living.' There are other events that do not.

Although in this analysis, the living part of the cosmos is very tiny in relation to the whole, the foundational entities of the whole are organisms. There is no 'matter' as that is usually understood metaphysically. The rudiments of what develops into life are everywhere.

The Presence of the Past, Novelty Potentiality, and the Cosmos as Life Itself

I turn now to the question of the cosmos as a whole. Is it alive? I have argued that the entities of which it is ultimately composed are events that are not material. These events have characteristics lacking in the idea of matter. I have recognized that the entities composed of these events may be as a whole lacking in any dimension of living. The stone and the machine are examples. Hence, the question about the whole is by no means settled by what we have learned about its ultimate constituents. The whole not only is ultimately composed of non-material entities. It also contains the whole gamut of organisms. But the whole as such may be simply the multiplicity of the entities that compose it or an inert and inanimate unity of those parts.

However, it is my conviction that our almost inescapable actual beliefs attribute to the whole characteristics and functions that it could not have if it were like a stone or a machine. I will try to persuade you that this is so, or at least that you act and live as if this were so. To do so, I must ask you to imagine what you would experience if the whole were simply material.

Consider first what you think about the status of the past. Does the past "exist"? If the universe as a whole were more like a stone or a clock than an organism or a living thing, the present could not include the past. What exists, then, is only the present. Statements about the past are neither true nor false. Mr. X neither committed the murder for which he is being tried nor failed to do so. That is past and therefore does not exist. One statement about it is as empty of meaning as another. To say that someone promised you something yesterday is equally meaningless. There is nothing for which to be grateful to your parents other than what they are doing in this moment.

Now I do not suppose that anyone lives or acts or thinks in this way. For one thing, we all have memories. The first rejection of this view is to point to memory. Our memories give us information about what happened in the past. We can live our lives now in terms of those memories. They are part of the present.

This is certainly true. Our lives are deeply affected by our memories. But what are memories? We experience them as connecting us to something that actually happened. When did it happen? In the past. Where is the past? We are positing that it is nowhere, that it does not exist. The memory is in the present. But if the past as past no longer exists, the memory cannot be distinguished from imagination in the present. The idea that what we call memories are about the past persists. We are sure that memory is different from imagination. The problem is that the assumption of the nonexistence of the past undercuts any possibility of a significant difference. Just as it is meaningless to ask whether your imagination is true or false; so, if the past does not exist, statements supposedly about it cannot be true or false. But we all act as though they are.

Scientists develop theories about the past. One of them is that there was a Big Bang. They have launched an experiment to determine whether this theory is true and to answer some questions about what happened. They do not suppose that the experiment will cause the Big Bang to have happened. They do not suppose that the experiment will make it true. It is already either true or false even though no one has any memory one way or another. In other words, even though the materialist metaphysics to which they typically subscribe has no place for a past, they act as though they believe that the past in all its details is still what it was. They expect to learn more about it.

I hope that you will all agree that we all think and act on the assumption that the past now exists in some way, a way that does not depend on our experience of it. Indeed, what exists in the present is largely the result of what happened in the past. Efficient causes are all in the past about whose existence we are reflecting. No one really doubts the existence of the past. But materialism has no place for it.

I am arguing that the past of the cosmos exists in full rich detail independently of our knowledge. It seems that the whole includes the whole past. If the whole were like a stone or a machine, or if it were at any moment simply what it is in that present, it would not do so. To encompass the past within its present, the cosmos as a whole must be much more than the multiplicity of its parts. It must include something like a perfect memory, only more than a memory, it must include the past as the past, still efficacious in the present.

Does this mean that we all believe the cosmos is alive? Not in the strict sense. We believe it has a property in an utterly superior way that our memory participates in fragmentarily. Possessing a memory is a remarkable property of some living things, especially animals. Including the entire past so as to keep it all effective in the present requires some kind of super-life.

I believe there is another widespread belief that I would find difficult not to hold that points to there being much more in the cosmos than the collection of individual events. In general, the scientific worldview, although it cannot explain the efficacy of the past in the present, regards that efficacy as fully determinative of the present. A materialistic world ends up, almost necessarily, as a determinist one. But our normal experience is that we sometimes make important decisions, and many of us feel that we are making minor decisions all the time. Why do scientists reject the idea of decision? It does not fit their worldview. The reality of a decision requires that in the moment, in addition to the causality of the past, there are alternative possibilities, a world of potentiality. The idea of matter does not have any place for alternative possibilities. But the novelty associated with life is not conceivable without them.

If I do not worry about being scientific, and simply describe my experience, a great deal of it feels determined by my personal past and other elements in the environment, including other people. But I do not feel that just how I respond is completely determined by all of this. Sometimes I feel that I can either speak or be silent, for example. My decision may affect the course of the discussion. If the scientist tells me that what has happened in the past determined that I would speak or that I would remain silent, I know that he may be right. My sense of having some personal responsibility for what I did may be an illusion. But in the moment of decision, it seems that what I do is not determined until I determine it.

Most parents feel that their children as they grow older are responsible for some aspects of what they do. Making distinctions between what their children choose and what to blame on circumstances and others is important for encouraging moral behavior. None of this makes any sense if all behavior and thinking and "choosing" are predetermined. The legal system as a whole makes no sense if there is no such thing as responsibility.

The great majority of scientists, when not engaged in science or teaching science, assume they have some responsibility for their choices and also bring up their children to think this way. If the sense that we have some responsibility for what we do, that there are sometimes real alternatives, is so deep that virtually all societies ground some of their organization upon it, should we deny our experience in order to be scientific? I do not believe this is actually possible, and I see no reason to try.

The reason scientists make these judgments that are in conflict with human experience is that their metaphysics has no place for much of human experience. It tends to claim that everything fits the materialist paradigm and to deny the reality of what does not. We are asking whether reality has dimensions that do not fit. The answer is that it does.

A metaphysics that allows for the truth of what we know experientially is better than one that does not. That means that a metaphysics that allows for possible alternatives is better than one that does not. One that allows for such alternatives need not limit its applicability to science, and it may enable scientists to explain phenomena that are inexplicable with their preferred metaphysics. This preferable metaphysics informs us that the cosmos is such that there are unactualized potentialities. The novelty that defines life is grounded on these potentialities. The cosmos as a whole includes the fundamental features that make life possible and actual. It may not be alive in the same way that we are, but it grounds life by its eternal inclusion of novel potentiality. Once again it is characterized by super-life. We may suggest this by capitalizing 'Life.' We are alive. The cosmos, or a crucial dimension of the cosmos is Life Itself.

Once again, if the characteristics of *living* are taken from vegetable and animal organisms on this planet and limited to those, then, those plants and animals are the only examples we know of living organisms. The plasma and quarks are not living, and neither is the cosmos. If one insists on a specific definition of 'life' or 'living,' one will have to invent other terms to describe the diversity of ways in which things transcend inert materiality. We might call the former pre-life and the latter *Life Itself*.

The Self-Enrichment of Cosmic Life in Its Parts as Our Personal Guidance

Our third topic is the question of whether the cosmos favors the enrichment of life in its parts. I think so. Indeed, there is now a lot of evidence that the cosmos is ordered to the end of making complex forms of life, such as the plants and animals on this planet, possible. So strong is this evidence, that the scientists who want to explain the presence of life on this planet as a matter of chance have affirmed that there must be thousands of universes of which only ours gives rise to this kind of life.

What scientists have found is that our cosmos is finely tuned to make life possible. There are many details about nature that appear to be arbitrary. They could just as well be different. Yet it turns out that if any of them were different, the kind of life that has so richly developed on this planet would not have been possible. If the chance of all of these being just what they are, is one in a hundred or even a thousand, common sense would suggest that there was a cosmic decision in favor of life.

Materialists do not believe in decisions at any level, certainly not at the cosmic one. But when the reality is so improbable on the basis of chance, they feel the need to come up with an alternative theory. The only one of which I am aware is that there are or have been many universes in which the life-favoring characteristics are absent. By chance they characterize our cosmos. So far as I know, this theory has not been worked out in detail. In my judgment, the more fully it is developed the less plausible it will be. Since the only argument in its favor is that it avoids a cosmic decision, and since I have already explained my objection to avoiding human decisions, I think scientists are far too committed to a metaphysics that cannot explain many well-known phenomena and flies in the face of common sense. Unless some evidence or argument is given against cosmic decision, we are justified in stating that the cosmos favors life.

We may, of course, ask how the cosmos acts. One account develops out of what was described in the previous section. The cosmos includes potentiality as well as actuality. The ordering of potentiality affects what becomes actual. The discovery that the universe is finely tuned to make life possible is explained as the ordering of potentiality.

The ordering of potentiality establishes, or participates in establishing, what used to be called natural laws. It also grounds the sense of responsibility for our choices. Our experience suggests that it offers us an ideal possibility for the situation in which we find ourselves. It does not compel us to appropriate it. Our decision may bring us close to that ideal, but we often "miss the mark." Many of us find ourselves drawn toward actions that we resist taking.

When we examine the potentials which we feel drawn to actualize, on the whole we can consider them life-enhancing. They enhance our own lives and those of others. Occasionally they may call for personal sacrifice for a greater good. We can generalize by saying that they embody the aim of the cosmos at the enhancement of life. They can be distinguished from the internalization of conventional teaching even though that is also often on the side of life.

I have stuck to types of experience that are very common, almost inescapable. However, many of us have also been blessed by experiences of being loved, or called, or guided in dramatic ways. Such experiences are often called religious. It is hard to believe that the cosmos can take such personal interest in individuals. But those of us who have had such experiences cannot doubt their authenticity and personal character. What we encounter cares about our lives and how we live them. We are called on to join the cosmic struggle for life, both human life and all life.

John B. Cobb, Jr. (USA) is one of the most important ecological and transdisciplinary thinkers of our time. He was a professor at Claremont School of Theology and Claremont Graduate University, and a visiting professor at prestigious universities in the United States, Germany, and Japan. In a career spanning more than seven decades, he co-founded many prestigious institutions and received numerous awards for his work.

Admitting Our Inalienable Links with the Cosmos

Chandra Wickramasinghe

Abstract This chapter traces the progress of studies spanning many scientific disciplines and more than forty years of research. They all converge on the conclusion that life is a cosmic phenomenon. A comprehensive perspective on this field of scientific and philosophical knowledge is provided in twelve basic steps. Starting with the fact of the complex organic nature of interstellar dust, the author reflects on the implications of accepting the concept of cosmic life. He also considers the subject of exoplanets and the search for extraterrestrial life. The author then turns to the panspermia theory, which he has been instrumental in reviving and developing significantly since the 1970s. He argues for the cosmic origin of life using probability calculations and then discusses the direct spectroscopic evidence and other geological evidence for panspermia. The distribution of habitable exoplanets and the possibility of interplanetary transfer of life forms are discussed. This is complemented by the history of the research into meteorites, micrometeorites and cometary bacteria. Recent relevant studies on the controversial Polonnaruwa Meteorite are considered. With regard to panspermia, stratospheric sampling and the idea of diseases and pandemics from space are presented as important areas of existing and future research.

Keywords Panspermia · Cosmic life · Comets · Interstellar dust · Microfossils · Meteorites · Societal constraints

C. Wickramasinghe (✉)
Buckingham Centre for Astrobiology, University of Buckingham, Buckingham, UK
e-mail: ncwick@gmail.com

Centre for Astrobiology, University of Ruhuna, Matara, Sri Lanka

National Institute of Fundamental Studies, Kandy, Sri Lanka

Institute for the Study of Panspermia and Astroeconomics, Gifu, Japan

D. Bartosch et al. (eds.), *Towards a Philosophy of Cosmic Life*,
https://doi.org/10.1007/978-981-99-2131-7_3

Introduction

> An error does not become truth by reason of multiplied propagation,
>
> nor does truth become error because nobody sees it.
>
> Truth stands, even if there be no public support.
>
> It is self-sustained.
>
> Mahatma Gandhi

In most ancient cultures throughout the world, comets were regarded with reverence as well as awe. They were thought of as bringers of pestilence and death on the one hand, and of life and fortune on the other (Yeomans 1991, 78, etc.; Temple 2007). In the present era, however, we tend to dismiss all such views as ignorant superstitions. When, together with Fred Hoyle, the present writer "revisited" comets from the late 1970s onwards and came to argue that aspects of such ancient superstitions might well be fortuitously true (Hoyle and Wickramasinghe 1978b, 1982a, 227, 1985, 1986, 509, 2000), we were often fiercely criticised. An unwritten rule of Western modern science appears to be that no vestige of a discarded ancient belief or sets of beliefs could be revisited, let alone revived.

From the 1970s onwards, the present author, in collaboration with Fred Hoyle and later with other collaborators, began to assemble a vast body of data and evidence to support panspermia from astronomy, geology as well as biology. New data and new facts continue to provide ample verification of prior predictions with ever more compelling evidence pointing to the *inevitability* of panspermia, as opposed to spontaneous generation, as the mode of origin and propagation of life throughout the universe. We have argued from the 1980s through to the present day that comets are incubators and distributors of life in the form of bacteria and viruses throughout the universe. Today, whilst comets are conceded at long last as the repositories of complex organic molecules that may have contributed to spontaneous generation of life on the Earth, their role as carriers of life, despite a great deal of contrary evidence, is still fiercely resisted in many quarters.

Over the past century or so, astronomy has established beyond a shadow of doubt that the universe and the Earth are inextricably linked at least in one regard. All the chemical elements that are present everywhere around us—the C, N, O in our bodies, the silicon in the rocks, the iron and other chemical elements that make up the huge multiplicity of objects and artefacts around us—were synthesised in nuclear reactions that take place in the deep interiors of stars (Burbidge et al. 1957). Although this proposition was disputed for a while, it is now part and parcel of orthodox everyday accepted science. Over the past half a century or so, it has been slowly emerging that our ancestry as living biological entities also resides in the stars (Hoyle and Wickramasinghe 2000; Wickramasinghe 2010, 119–129, 2015).

Interstellar Dust

When the present author first came to investigate the nature of interstellar dust in 1960, the vast amounts of dust in the Milky Way were thought to be comprised of micrometre sizes ice particles, similar to the ice grains found in the cumulous clouds in the Earth's atmosphere. A re-examination of this earlier work soon led to the conclusion that this model of cosmic dust was wrong, and instead the theory was proposed that cosmic dust was made up mainly of the element carbon (Hoyle and Wickramasinghe 1978a, b). The precise form of the carbon in cosmic dust was gradually unravelled by examining the light from distant stars that traversed through the dark clouds leaving spectroscopic signatures of their chemical make-up.

By the mid-1970s there was clear astronomical evidence for the widespread occurrence in the galaxy of organic molecules (Hoyle and Wickramasinghe 2000). At about this time, infrared observations of interstellar dust were beginning to show spectral features in the mid- and near-infrared which could not be easily reconciled with a combination of inorganic silicates and water ice, as was hitherto believed. But this evidence was vigorously disputed by our critics for a long time. Fred Hoyle and the present writer spent nearly five years of our professional lives modelling this data with biopolymers—polysaccharides, polyaromatic hydrocarbons in particular. And our pioneering work has at last been vindicated by a flood of data coming in from space telescopes and space probes (Hoyle and Wickramasinghe 1976, 45–46, 1977, 323–324, 1978a, 489–505). The complex organic nature of the dust in interstellar space as well as in comets and meteorites is now beyond dispute.

The manner in which some of these organic molecules might relate to life and biology, however, remains a matter of controversy—one fuelled by cultural constraints rather than science. The question relates to whether these biologically relevant molecules represent steps towards life, prebiotic evolution, or whether they are the products of biological degradation—the detritus of life. The overwhelming bulk of the organic material found on Earth is the result of the decay of biology. So, the question we need to ask is this: Why is it not the same for the organics in space?

The currently fashionable idea is that the chemical building blocks of life may be delivered to a primitive Earth to form a "primordial soup" from which life thereafter is supposed to emerge through chemistry even against manifestly insuperable odds. My own position has been markedly different and has been guided by a long succession of discoveries in astronomy, geology and biology. The emerging view, which is still being denied, is that life is a truly cosmic phenomenon and arose on Earth following the delivery of bacteria and genetic components in the form of viruses by means of comets. On this picture, the patterns of life's evolution witnessed on Earth must be reproduced on countless other planetary abodes—exoplanetary systems such as been shown to exist in their hundreds of millions in our galaxy alone.

Implications of Accepting the Concept of Cosmic Life

We live today in a world that is firmly centred on an anthropocentric and overwhelmingly Eurocentric viewpoint. The assertion of our self-importance as a species and as a superior tribe of humans has ramifications even in relation to our attitude to the non-human living world, as well as on our perceived right to exploit the latter in any manner we might choose fit. This is reflected in our reaction to such issues as climate change and environmental degradation resulting from our conduct.

If it can be firmly established that we—as sentient living entities—are not alone in the universe, the implications for humanity will be profound. It could be even more important if we know that alien life in the form of microbes—bacteria and viruses—exists in our midst even now and continually rains down on our planet. Such microbes may sometimes cause devastating pandemics of disease as happened very recently with the COVID-19 pandemic (Hoyle and Wickramasinghe 1979); but more positively they have the potential to augment our genomes and over long intervals of time unravel an ever-changing panorama of cosmic life (Hoyle and Wickramasinghe 1982b). In either case—whether as alien microbes in our vicinity, or alien intelligence on distant planets—the ultimate acceptance of the emerging facts of cosmic life will mark an important turning point in human history.

But how did life arise in the first instance? Not just on Earth, but anywhere in the universe? Does life emerge spontaneously on every Earth-like planet by processes involving well-attested laws of physics and chemistry from simpler chemical units such as nucleobases and amino acids, even if all these can be delivered from space? Or did the first-ever origin of life involve an extraordinary unrepeatable event involving stupendous volumes of the universe in the most distant cosmological past? These age-old questions have recently acquired a new sense of urgency. As I have indicated, the answers stare us in the face, but we tend to turn away because they are unpalatable and demand a humiliating diminution of our perceived status in the cosmos.

Exoplanets and the Search for Evidence of Extraterrestrial Life

The primary requirement for the emergence of life, and subsequently of creatures like ourselves, would be for the existence of rocky planets with water and an atmosphere generally similar to that of Earth. Mayor and Queloz have discovered definite evidence of planets outside our solar system (1995). The first of these so-called exoplanets orbited a star some fifty light years away in the constellation of Pegasus; it was a giant planet with a mass similar to Jupiter but located too close to its parent star for any life to be possible. In 2009, NASA launched its orbiting Kepler telescope, which was specifically designed to discover planets that are the size of Earth. The detection process involved tracking down minute blinks (dimming) in the star's light when a planet transited periodically in front of it during its orbit. At present, more than 5000 definite as well as probable detections of habitable planets have been made within only a very small sampling volume of our Milky Way. Most of these planets orbit red dwarf stars that are on the average twice the age of our Sun. Extrapolating from the sample of present detections, the estimated total number of habitable planets in our Milky Way galaxy is reckoned to be in excess of 100 billion (Kopparapu 2013). On many of these planets, one might speculate, life may have begun, evolved and perhaps long since disappeared.

Another related enterprise that has captured the news recently is the search for extraterrestrial intelligence using arrays of radio telescopes to scan the skies for evidence of intelligent signals. Over half a century ago, Philip Morrison and Giuseppe Cocconi first drew attention to the possibility of searching the microwave spectrum of cosmic sources for intelligent signals and suggested particular frequencies as well as a set of potential targets. The SETI program (Search for Extraterrestrial Intelligence) began tentatively in 1960 and was first supported by NASA, and later by a host of private or semi-private entrepreneurs. With the exception of a single brief and mysterious "Wow!" signal discovered in August 1977, there has been a deathly silence across all of the prospective sources that have been scanned.

There could be a case for saying that the lack of progress in this venture was the result of organisations like NASA backing off. This may have been the thinking behind Russian billionaire Yuri Milner's 100 million-dollar SETI initiative that was recently announced with much pomp. Buying more telescope time, increasing the range of wavelengths, enhancing detector sensitivity and extending sky coverage have been argued as prerequisites if a breakthrough within a decade is to be achieved. It is also possible that we have the wrong strategy altogether in our SETI endeavours. Predrag Slijepcevic and the present writer have recently argued that we may be looking in the wrong places for SETI (Slijepcevic and Wickramasinghe 2021). What if the ET-messages are not in the form of electromagnetic radiation but coded messages in the DNA and RNA of viruses that already arrive on our planet? One way or another, a positive result from SETI would be contingent on the emergence and dispersal of primitive life capable of evolving into intelligent creatures such as ourselves.

Panspermia

How common is the phenomenon of life in the universe? The answer to this question lies in the concept of panspermia—or cosmic life. The idea that microbial life springs up de novo on billions of Earth-like habitable planets is an unproven, and most likely erroneous, proposition. Such a belief is an extension of the canonical "primordial soup theory" for life's beginnings on Earth, which is a dogma with no hard evidence to support it. The dogma, in turn, can be traced back to the Greek philosopher Aristotle (384–322 BCE), who proposed that life arose spontaneously from inorganic, inanimate matter whenever favourable conditions prevailed—for example, fireflies from a mixture of warm earth and morning dew! Unfortunately, because of his stature as a philosopher, Aristotle's views prevailed and still prevail almost to the present day. Experimental demonstrations against the validity of spontaneous generation have consistently been ignored in particular since the 19th-century experiments of Louis Pasteur, showing that life at a microbial level is always derived from pre-existing microbial life.

If there was a deep principle of nature that drove inorganic systems towards the emergence of primitive life—bacteria rather than fireflies—the evidence for this would have long since been discovered in the laboratory. With the calculations in the next section showing grotesquely low a priori probabilities for the transition from non-life to life, only two options remain. The origin of life was an extremely improbable event that occurred on Earth (because we are here!) but will effectively not be reproduced elsewhere. In that case we would indeed be hopelessly alone. Or, a very much vaster cosmic system than was available on Earth and a very much longer timescale were involved in an initial origination event, after which life was transferred to Earth and elsewhere by processes that the present writer and the late Sir Fred Hoyle have proposed many years ago—cometary panspermia. *The essence of our theory of cosmic life is that the entire galaxy—perhaps the entire universe—is one single connected biosphere.* Physical transfer involving exchange of meteorites, comets and dust with an intermingling of genes on a cosmological scale is far more probable than independent life origination events that are assumed in conventional science.

The Mind-Blowing Improbability of Life

The operation of a living system depends on thousands of chemical reactions taking place within a membrane-bound cellular structure—for example, a bacterial cell. Such reactions, grouped into metabolic pathways, have the ability to harness chemical energy from the surrounding medium in a series of very small steps: transporting small molecules into cells, building long chain biopolymers of various sorts, and ultimately making copies of themselves while possessing a capacity to further evolve. Batteries of enzymes, composed of chains of amino acids, play a crucial role as catalysts precisely controlling the rates of chemical reactions. Without enzymes, and the specific arrangements of amino acids within the enzymes, there could be no life.

In present-day biology, the precise "information" contained in enzymes—the arrangements of amino acids into folded chains—is transmitted by way of the coded ordering of the four nucleotide bases (A, T, G, C) in DNA. In a hypothetical RNA world that some biologists think may have predated the DNA–protein world, RNA is posited to serve a dual role as both enzyme and transmitter of genetic information. If a few such ribozymes are regarded as precursors to all life, one could attempt to make an estimate of the probability of the assembly of a simple ribozyme composed of 300 bases. This probability turns out to be 1 in 4^{300}, which is equivalent to 1 in 10^{180}, which can hardly be supposed to happen even once in the entire 13.9-billion-year history of the universe. And this is just for a single enzyme. In the simplest known bacterium *Mycoplasma genitalium* with some 500 genes coding for enzymes, the improbability escalates to a super-astronomical scale (Hoyle and Wickramasinghe 1982b, 2000; Wickramasinghe 2015).

Over the past fifty years, the most advanced laboratories in the world have continued to test the theory of spontaneous generation by attempting to synthesise an autonomous bacterial cell. The failure thus far to witness any trend whatsoever toward the emergence of a living system is often attributed to the infinitesimal scale of the laboratory system when compared with the terrestrial setting in which life is thought to have arisen. Yet, if we move from the laboratory flask to the oceans of the Earth, we gain in volume only a factor of about 10^{20}, and in time, from weeks in the laboratory to half a billion years, the gain is an additional factor of 10^{10}. In the probability calculation for the single ribozyme, we therefore gain a factor of 10^{30} in all, which reduces the improbability factor from 1 in 10^{180} to 1 in 10^{150}. A grotesque improbability remains, and that just for a single enzyme!

On this basis, it is very difficult to avoid the conclusion that the emergence of the first evolvable autonomous cellular life form was almost certainly a unique event in the cosmos. If this did indeed happen on Earth for the first time, as conventional Eurocentric, implicitly Neo-Aristotelian "wisdom" tends to assert, it must logically be regarded as a near-miraculous event that would not be repeated elsewhere, let alone in any laboratory simulation of the process. To overcome improbabilities on the scale that is involved here, it stands to reason that one would gain immensely by going for the biggest system available—manifestly, the universe as a whole.

The argument that panspermia must be rejected because it merely transfers the problem of origin from Earth to another setting has no logical basis whatsoever. The question of whether life started de novo on Earth or was introduced from the wider universe is a fully scientific and precise question that merits investigation—one that is open to testing and verification. The invocation of Occam's razor to exclude a discussion of such matters is unfortunate and merely an excuse for keeping scientific discussion within the strict bounds of what is considered *de rigueur*. It is strikingly reminiscent of the kind of restrictions that stifled science during the Middle Ages.

Bold assertions being made that panspermia is invalid because viable bacteria and viruses cannot be transported in viable form across astronomical distances is a striking example of the wrong and inappropriate use of Occam's razor. Despite claims to the contrary, the weight of scientific evidence does indeed favour survival of bacteria under space conditions to at least to the extent that makes viable interstellar transfers of life inevitable. We do not require more than one in 10^{24} iterant microorganisms to survive until they become incorporated in a planet-forming or comet-forming event in which a new cycle of exponential amplification occurs; a few viable microbes then turning into trillions.

This exceedingly modest requirement of microbial survival would be impossible to violate particularly for freeze-dried microorganisms embedded within particles of interstellar dust. The vast majority of bacteria in interstellar space does not and need not persist in viable state. Interstellar clouds would thus be filled overwhelmingly with the detritus of life which takes the form of genetic fragments that could include viruses as well as a wide range of organic molecules. The very basic idea is that of a cycle: Biomaterial is developing in the contexts of stars, planets and comets, and amplified biomaterial finds its way back into interstellar space. Bacteria expelled from a planetary system are amplified in the warm radioactively heated interiors of comets and thrown back into interstellar clouds, which become the seeding ground for further developments in contexts of cometary and planetary evolution of life forms (Steele et al. 2019, 24, also Fig. 8).

Direct Spectroscopic Proof

It would be false modesty to maintain that the late Sir Fred Hoyle and the present writer were not directly responsible in large part for the revival of panspermia in the form it is being discussed in 2023. Our proposal was that trillions of cometary bodies in the cosmos serve as the long-term reservoirs and repositories of cosmic biology, evidence for which shows up in comets as well as in the all-pervasive interstellar dust in the universe as we have already noted.

The next key development that took place in 1986 with the return of Halley's Comet after seventy-six years. This was the first comet that came under close scrutiny after the dawn of the Space Age. The popular view at the time that comets were dirty snowballs (Whipple's dirty snowball comet) came to be demolished fairly quickly by observations that were made in April 1986 using satellites as well as ground-based telescopes. Before the actual rendezvous of the Giotto Space Probe with Comet Halley in April 1986, nobody had ever seen the nucleus of a comet—only their spectacular tails were seen and photographed, and people had to theorize and conjecture as to the existence and the nature of the comet's nucleus or core. Ahead of the observations of 1986 being made, and based our organic comet model, Fred Hoyle and the present author made the firm prediction that when it was heated in sunlight near the perihelion (closest approach to the Sun), it would boil off bacterial dust and turn into a black coal-like body. In a dramatic way, this prediction was verified. Moreover, an infrared spectrum of the comet's dust tail was found to match the spectrum of a bacterium to an amazing degree of precision (Hoyle and Wickramasinghe 1986; Wickramasinghe 2010). For respective figures based on the data, see the author's published drawings in Steele et al. (2018, 8, Fig. 2, left panel) and Steele et al. (2019, 23, Fig. 6).

Since 1986, a wealth of data has been accumulated relating to other comets, from ground-based telescopes as well as space probes, and these have been extensively reviewed in a long series of scientific papers and books. In no case could we find any contradiction with our biological model of comet dust, although astronomers are still loath to admit the fact. The latest Rosetta space probe to the comet 67P/CG has generated a wealth of data that can only be explained satisfactorily on the basis of a biological model of the dust (Capaccioni et al. 2015). The surface reflectivity spectra of comet 67P/C-G compared with the transmittance curve measured for the bacterium *Escheria coli* shows the close consistency between the surface properties of the comet 67P/C-G and the spectrum of a desiccated bacterial sample (Capaccioni et al. 2015, 3, Fig. 1, right panel; see also Steele et al. [2019, 24, Fig. 7, right panel]).

One aspect of cometary behaviour that might serve to decide critically between a biological and non-biological model was already stressed by Wickramasinghe et al. more than twenty-five years ago (1996). This is the profuse emission of gas and dust occurring for some comets in the cold depths of space far beyond the orbit of Jupiter. At such distances, both thermal evaporation and surface detonation of chemical processes would be ruled out. The first such phenomenon was discovered for Comet Hale-Bopp erupting at a distance of 6.5 AU (astronomical units) in 1983/84 long before it had reached perihelion in April 1997. More recently in 2015, Comet Lovejoy, erupting much closer to the Sun at 1 AU distance, emitted vast amounts of a sugar and ethyl alcohol, which are natural products of fermentation (Biver et al. 2015). The ethanol and sugar thus released amounted to an alcohol equivalent of 500 bottles of wine every second—plausibly the products of fermentation in subsurface liquid domains of the comet!

More recently, the discovery of a giant comet (C/2014 UN271) some 100 kilometres in diameter at a distance of 29 AU in October 2014 and the later discovery in September 2021 of a dramatic brightening episode offers a further opportunity for verifying the predictions of fermentation processes in a "biological" comet (Wickramasinghe 2022a). The eruptions of the comet at a heliocentric distance of 20 AU (two thirds of the distance from the Sun to Neptune) can only be plausibly explained as due to high pressure venting of the products of microbial metabolism in radioactively heated subsurface lakes. It is currently (in January 2022) active even further out near the very edge of our planetary system. The standard non-biological model of comets is woefully inadequate to account for eruptions at such large distances from the Sun, where surface temperatures are as low as 60 K (Wickramasinghe 2022a). Despite all this strongly supportive data, there appears to be a concerted effort on the part of a majority of scientists and science journalists to assert that although life-related organic molecules have been found in large quantity in comets and also asteroids they cannot be interpreted as signifying life.

Geological Evidence Supporting Panspermia

Three decades ago, the earliest evidence for microbial life in the geological record was thought to be in the form of cyanobacteria-like fossils dating back to 3.5 Ga (billion years) ago. From the time of formation of a stable crust on the Earth 4.1 billion years ago, following an episode of violent impacts with comets (the Hadean Epoch), there seemed to be available an 800 million years timespan during which the canonical Haldane-Oparin primordial soup and the spontaneous generation of life may have arguably developed. Very recent discoveries, however, have shown that this time interval has been effectively closed. Ancient rocks laid down 4.1 to 4.2 billion years ago, which belong to a geological outcrop in the Jack Hills region of Western Australia, have been found to contain micron-sized graphite spheres with an isotopic signature of biogenic carbon locked away within crystals of zirconium (Bell et al. 2015). Such signatures of fossil bacteria were deposited at a time when the collisions of the Earth with comets and asteroids were happening at a relentless pace.

This latter discovery in my view puts paid to the possibility of any primordial soup brewing on Earth, because these fossils were deposited at a time when the planet was being relentlessly bombarded by comets and meteorites.

The requirement now, on the basis of orthodox thinking, is that an essentially instantaneous transformation of non-living organic matter to bacterial life took place, a proposition that strains credibility of earthbound abiogenesis to its utmost limit. A far more plausible proposition is that fully-developed microorganisms arrived at the Earth via impacting comets, and these became carbonized and trapped within ancient rocks.

Distribution of Habitable Exoplanets and Exchanges of Living Entities

We have already pointed out that more than 5000 definite as well as probable detections of habitable planets have been made within only a very small sampling volume of our Milky Way (see also Kopparapu 2013). Most of these planets orbit red dwarf stars that are on the average twice the age of the Sun. Extrapolating from the sample of detections in our local vicinity the estimated total number of habitable planets in the entire Milky Way galaxy is reckoned to be in excess of 100 billion. *Proxima Centauri b* (also called *Proxima b*) is closest habitable exoplanet orbiting the red dwarf star Alpha Centauri at a distance of some 4.2 light years from the Sun.

Whilst comets could supply a source of primitive life (bacteria, viruses and genes) to interstellar clouds and thence to new planetary systems and embryonic exoplanets, the genetic products of evolved life (local evolution) could also be disseminated on a galaxy-wide scale (Wallis and Wickramasinghe 2004). At the present time, our planetary system, which is surrounded by an extended halo of some 100 billion comets (the Oort Cloud) and replete with microbial content as we have seen, moves around the centre of the galaxy with a period of 230 My. Every 40 million years, on the average, this cloud of comets in our solar system becomes gravitationally perturbed due to the close passage of a giant molecular cloud. Such gravitational interactions lead to hundreds of comets from the Oort Cloud being thrown into the inner regions of our planetary system, some to collide with the Earth. Such collisions do not only cause extinctions of species (as one impact surely did 65 million years ago, killing the dinosaurs), but they could also result in the expulsion of surface material back into deep space (Wallis and Wickramasinghe 2004).

A mechanism can thus be identified for the genes of evolved Earth life to be transferred to alien habitable exoplanets. A fraction of the Earth debris so expelled survives shock-heating and could be laden with viable microbial ecologies as well as genes of evolved life. Such life-bearing material from the Earth could reach newly-forming planetary systems in the passing molecular cloud within a million years of the ejection event. A habitable exoplanet could then become "infected" with terrestrial microorganisms and terrestrial genes that can contribute to the process of local biological evolution.

Once life has got started and evolved on an alien planet or planets of a new system, the same process can be repeated (via comet collisions) transferring a new compliment of genetic material carrying local evolutionary "experience" to other molecular clouds and other nascent planetary systems. If every life-bearing planet transfers genes in this way to more than one other planetary system (say 1.1 on the average), with a characteristic time of 40 My then the number of seeded planets after 9 billion years (lifetime of the galaxy) is $(1.1)^{9000/40} \sim 2 \times 10^9$. Such a large number of "infected" planets illustrates that Darwinian evolution, involving horizontal gene transfers, must operate not only on Earth or within the confines of our solar system but on a truly galactic scale. Life throughout the galaxy on this picture would inevitably constitute a single connected cosmic biosphere.

Meteorites, Micrometeorites and Cometary Bacteria

When a life-bearing comet makes its repeated orbits around the Sun, its volatile material with an entrained component of bacteria and viruses are progressively peeled away. Carbonaceous chondrites that occasionally fall to Earth represent fragments of comets denuded of volatiles but retaining a residue of silicates and more robust organic structures, possibly fossilised microbial forms.

The topic of microfossils in carbonaceous chondrites has sparked bitter controversy ever since they were first discovered in the 1960s by Claus et al. (1963). Since carbonaceous chondrites are generally believed to be derived from comets, the discovery of fossilised life forms in carbonaceous meteorites would provide strong *prima facie* evidence in support of the theory of life in comets and cometary panspermia. However, some contamination was discovered in early studies, and claims that *all* micro structures (so-called organised elements) discovered within meteorites were either artefacts or contaminants led to a general rejection of meteoritic microfossil identifications in the 1960s.

The situation began to change in the 1980s when Hans Dieter Pflug discovered a profusion of microfossils in ultrathin sections prepared from the Murchison meteorite, a carbonaceous chondrite that fell in Australia on September 28, 1969 (Pflug 1984). The contaminant-free experimental method adopted by Pflug appeared to be beyond reproach.

Pflug's discoveries, made after we had launched the theory of cometary panspermia, made it now very difficult logically to reject the fossil identification—but unfortunately they were. The hostility to Pflug's work was spear-headed by those who wished to maintain the orthodox unproven theory of abiogenesis on the Earth. More recent work by Hoover (1997, 2005; etc.) has provided even more striking evidence of the existence of microbial fossils in the Murchison meteorite and in several other carbonaceous meteorites. But the response continues to be to ignore this data—with the mantra that "extraordinary claims need extraordinary evidence".

The Disputed Polonnaruwa Meteorite: Recent Relevant Studies

Perhaps one of the most interesting recent developments in this area were related to a meteorite fall in Sri Lanka in 2012 (Wallis et al. 2013). Minutes after a fireball was seen by a large number of people in the skies over Sri Lanka on 29 December 2012, a large bolide evidently disintegrated and fell in the village of Araganwila, which is located a few miles away from the historic ancient city of Polonnaruwa. At the time of entry into the Earth's atmosphere, the parent body of this Polonnaruwa "meteorite" would have had most of its interior porous volume filled with water, volatile organics and possibly viable living cells. Pieces of this meteorite were found to be very porous, similar to the 162173 Ryugu asteroid (Hoover et al. 2020; Harris et al. 2002).

Fragments from a freshly cleaved interior surface of the Polonnaruwa meteorite were mounted on aluminium stubs and examined by Jamie Wallis under an environmental scanning electron microscope (Wallis et al. 2013). Images of the sample at low magnification displayed a wide range of structures that were distributed and enmeshed within a fine-grained matrix, examples of which are also shown in Wickramasinghe et al. (2013, Figs. 2, 3, 5, 7 and "Appendix"). These include fresh water and seawater diatoms and an extinct microbial fossil known as an acritarch. These structures were deeply ingrained in the rock matrix and the range of species that were found cannot be reasonably explained on the basis of terrestrial contamination. The space origin of these rocks was challenged on the grounds that they are far too porous for being of asteroidal or cometary origin.

This objection, however, has been shown to be false following the discoveries relating to the asteroid 162173 Ryugu that was found to have precisely such a property. It should also be recalled that earlier studies of the distribution of stable oxygen isotopes in the Polonnaruwa stones had left little doubt as to their cosmic origin (Wallis et al. 2013; Wickramasinghe et al. 2013). Furthermore, neutron activation studies on the stones carried out later by Hoover et al. (2020) have yielded even more compelling evidence of their likely space origin. From the distribution of heat-producing radioactive elements (Potassium, Thorium, Uranium and the rare earth elements) it has been asserted that the Polonnaruwa stones were consistent with a cometary/asteroidal origin where a parent body had once contained liquid water domains. Such domains in the parent body of these stones would have served as habitats for microorganisms, including the diatom species seen. Since these meteorites cannot be seen as any other than pieces of extinct cometary fragments, the idea of microbial life carried within comets would appear to be vindicated.

Stratospheric Sampling

Even more direct proof of ongoing panspermia would be achieved if we can demonstrate beyond doubt that microbial life is still arriving at the Earth. This could be best achieved by sampling the stratosphere, which was done under rigorously controlled conditions for the first time in 2001. In collaboration with the Indian Space Research Organisation, a group of UK scientists recovered, under strictly aseptic conditions, samples of stratospheric air from 41 kilometres from which positive detections of microorganisms were reported (Harris et al. 2002). The number of bacterial cells collected in a measured volume of the stratosphere at 41 kilometres led to an estimate of an in-fall rate over the whole Earth in the range 0.3 to 3 tonnes of microbes per day. This converts to some 20 to 200 million bacteria per square metre arriving from space every single day.

This truly vast number of infalling micro-organisms pales into insignificance when compared to bacteria and viruses normally resident at the Earth's surface, some of which could be occasionally lofted to heights of about three kilometres in upward air currents and subsequently brought down in mist and rain. The total flux of such bacteria and viruses falling back on the tops of the Sierra Nevada mountains was recently studied by Reche et al. (2018). The average downward flux of viruses from this height, mainly recycled from the ground, was estimated to be 800 million per square metre per day. If both the space-incident microorganisms and terrestrial microbes are mixed into this estimate their genetic differences could turn out to be difficult to detect. Efforts to do this, however, are clearly of paramount importance, *and this should be considered an international scientific priority.*

Recently microorganisms were also discovered on the outside of the International Space Station that orbits at 400 kilometres above the Earth by Grebennikova et al. (2018). There is no easy way to maintain that such microorganisms could have been lofted from the surface of the Earth, so strongly supportive evidence for panspermia continues to grow in many directions. More expensive and sophisticated investigations need to be carried out even on the samples collected so far if we are to prove beyond a shadow of doubt that these microbes are unequivocally alien. The sad truth is that funding for such vitally important experiments is well-nigh impossible to secure due to a deeply ingrained prejudice in favour of an Earth-centred theory of biology.

Diseases from Space and Pandemics

Studies of pandemics in history have led to discussions of the possibility of the space origin of viral diseases (Hoyle and Wickramasinghe 1979). One important piece of historic evidence that emerged 102 years ago relates to the influenza pandemic of 1918 and 1919. Reviewing all the available data relating to this pandemic, Dr. L. Weinstein wrote thus in the *New England Journal of Medicine*:

> Although person-to-person spread occurred in local areas, the disease appeared on the same day in widely separated parts of the world on the one hand, but on the other, took days to weeks to spread relatively short distances. It was detected in Boston and Bombay on the same day, but took three weeks before it reached New York City, despite the fact that there was considerable travel between the two cities. It was present for the first time at Joliet in the State of Illinois four weeks after it was first detected in Chicago, the distance between those areas being only 61 km (38 miles). (Weinstein 1976, 1058–1059)

In the context of an unknown or poorly defined origin of the current COVID-19 pandemic, and with the growing evidence in support of panspermia, a *panspermic* primary origin of this virus cannot be ruled out. Many aspects of the epidemiology of this new virus do indeed support the idea of a primary atmospheric fall-out followed by person-to-person spread; and the disentanglement of the two processes presents a continuing challenge to the world (Steele et al. 2021; Wickramasinghe et al. 2021).

In the absence of hard evidence that any virus is of earthly origin, a space origin of viruses in general cannot be ruled out ab initio. This is particularly the case since the total viral content of the Earth is truly astronomical and has by no means been fully charted. For example, a single litre of seawater collected in marine surface waters has been estimated to contain more than 100 billion viruses (Fuhrman 1999). The total viral content of the oceans is estimated to be 10^{30}; the vast majority of identified viruses are informationally rich bacterial phages, but an unknown component of other viruses is also included in this tally. This number is astoundingly super-astronomical and exceeds by more than a factor of a million the total number of stars in the entire observable universe which is $\sim 10^{24}$. This comparison of astronomically big numbers is a startling indication of a connection between life on Earth and the wider cosmos. For an unmistakable illustration of this striking structural similarity in the macrocosm and the microcosm of virology, one might think of region of the Milky Way, which is densely packed with stars, where average distance between foreground bright (nearer) stars is typically a trillion kilometres and, on the other hand, strikingly similar looking fluorescence images of drops of ocean water containing over a million of individual viruses, like the one provided by Fuhrman (1999, 543, Fig. 1).

The footprints of viruses in the genomes of all terrestrial life forms began to be unravelled at the turn of the millennium when whole genomes, including the human genome, came to be sequenced for the first time. One of the many surprises that followed was the discovery that what was once considered "junk DNA" actually had a viral origin. Perhaps as much as thirty percent of our DNA consists of retroviral sequences (endogenous retroviruses, ERVs)—RNA viruses that have reverse transcribed their RNA into DNA. These are the relics of ancient viruses of cosmic origin that may have actually contributed to our evolution over hundreds of millions of years—and they clearly disprove the criticism that cosmic viruses cannot be the cause of pandemic disease (Steele et al. 2021; Wickramasinghe et al. 2021).

On Earth, a continued arrival of new bacteria and viruses must have taken place from the time when microbial life was first introduced by comet impacts some 4.2 billion years ago. New viruses eventually succeed in becoming incorporated into the genomes of evolving lifeforms, this process continuing over billions of years of geological history. Such interaction of new viruses with evolved host species is not always harmonious, however. Throughout recorded history, we have clear evidence of a succession of pandemics of disease that have swept over the planet from time to time—the most recent pandemic of COVID-19 being a case in point.

Conclusion

In accordance with well-established conventions of scientific methodology, we can use a feedback loop to generate cycles of prediction, verification/falsification, and reaffirmation, so as to put any theory or hypothesis of such cosmic biology discussed in this chapter to an ever more stringent test. From the theory of a panspermic origin of planetary life predictions are deduced. These are put to the test by experiment or investigation. This leads to new observations and new data, which are used to provide foundations for panspermia by means of induction. Then the cycle is repeated. I need hardly reiterate that this ongoing process has indeed been used over many years, leading to a veritable catalogue of successes and confirmations that imply consistency with the hypothesis of cometary panspermia.

From a personal point of view, I find it curious to reflect that the spectroscopic identification of interstellar dust and molecules in space with biochemicals that I started in the 1970s has come into sharp focus in present times. Their biological relevance, once vehemently contested, is also now widely conceded, albeit to a very limited degree. The trend remains to assert without proof that we are witnessing the operation of prebiotic chemical evolution on a cosmic scale—a process we have discounted (earlier in this article) as inadmissible on probabilistic grounds.

If biological evolution and replication are regarded as the only secure relevant empirical facts—life always generates new life—this must be so even on a cosmic scale. Maintaining the premise of an Earth-centred origin of life, which some scientists still prefer to do, harks back to an outdated Aristotelian world view ("fireflies emerging from a mixture of warm earth and morning dew…") that has held sway for over two millennia. We conclude by reminding the reader that a very similar situation prevailed before the Ptolemaic Earth-centred solar system was finally abandoned. The historian of science Thomas Kuhn thus writes:

> The state of Ptolemaic [Earth-centred] astronomy was a scandal before Copernicus' announcement. (Kuhn 1996, 67, insertion CW)

> Given a particular discrepancy, astronomers were invariably able to eliminate it by making some particular adjustment in Ptolemy's system of compounded circles. But as time went on, a man looking at the net result of the normal research effort of many astronomers could observe that astronomy's complexity was increasing far more rapidly than its accuracy and that a discrepancy corrected in one place was likely to show up in another. (Kuhn 1996, 68)

Let us hope this will not be the case for the overwhelming evidence of cosmic biology in the year 2023 and that cultural obstacles to accepting the facts will be overcome (Wickramasinghe 2022b). *Conceding a long overdue paradigm shift from Earth-centric biology to cosmic life would have far-reaching implications that go beyond science*. With a radical revision of our place within the cosmos, our self-acclaimed supremacy as humans will change beyond all recognition. This could turn out as the only route to a sustainable future, where humans learn at last to live in peace amongst themselves and in harmony with the wider natural environment.

References

E.A. Bell, T. Patrick Boehnke, M. Harrison, W.L. Mao, Potentially biogenic carbon preserved in a 4.1 billion-year-old zircon. PNAS **112**(47), 14518–14521 (2015). https://doi.org/10.1073/pnas.1517557112

N. Biver, D. Bockelée-Morvan, R. Moreno et al., Ethyl alcohol and sugar in comet C/2014Q2 (Lovejoy). Sci. Adv. **1**(9), e1500863 (2015). https://doi.org/10.1126/sciadv.1500863

E.M. Burbidge, G.R. Burbidge, W.A. Fowler, F. Hoyle, Synthesis of the elements in stars. Rev. Mod. Phys. **29**(4), 547–650 (1957). https://doi.org/10.1103/RevModPhys.29.547

F. Capaccioni, A. Coradini, G. Filacchione et al., The organic-rich surface of comet 67P/Churyumov-Gerasimenko as seen by VIRTIS/Rosetta. Science **347**(6220), article aaa0628 (2015). https://doi.org/10.1126/science.aaa0628

G. Claus, B. Nagy, D.L. Europa, Further observations on the properties of the 'organized elements' in carbonaceous chondrites. Ann. N. Y. Acad. Sci. **108**, 580–605 (1963). https://doi.org/10.1111/j.1749-6632.1963.tb13409.x

J.A. Fuhrman, Marine viruses and their biogeochemical and ecological effects. Nature **399**, 541–548 (1999). https://doi.org/10.1038/21119

T.V. Grebennikova, A.V. Syroeshkin, E.V. Shubralova et al., The DNA of bacteria of the world ocean and the earth in cosmic dust at the international space station. Sci. World J. **2018**, article 7360147 (2018). https://doi.org/10.1155/2018/7360147

M.J. Harris, N.C. Wickramasinghe, D. Lloyd et al., Detection of living cells in stratospheric samples, in *Proceedings SPIE*, vol. 4495. Instruments, Methods, and Missions for Astrobiology, vol. IV (2002), pp. 192–198. https://doi.org/10.1117/12.454758

R.B. Hoover, Meteorites, microfossils, and exobiology, in *Proceedings SPIE*. Instruments, Methods, and Missions for the Investigation of Extraterrestrial Microorganisms, vol. 3111 (1997), pp. 115–136. https://doi.org/10.1117/12.278766

R.B. Hoover, Microfossils, biominerals, and chemical biomarkers in meteorites, in *Perspectives in Astrobiology*, ed. by R.B. Hoover, A.Y. Rozanov, R. Pape. NATO Science Series I: Life and Behavioral Sciences, vol. 366 (IOS Press, Amsterdam, 2005), pp. 43–65

R.B. Hoover, M. Frontasyeva, S. Pavlov, Epithermal neutron activation analysis of carbonaceous chondrites and the Polonnaruwa/Aralaganwila stones. Asp. Min. & Miner. Sci. **6**(1), article AMMS.000626 (2020). https://doi.org/10.31031/AMMS.2020.06.000626

F. Hoyle, N.C. Wickramasinghe, Primitive grain clumps and organic compounds in carbonaceous chondrites. Nature **264**, 45–46 (1976). https://doi.org/10.1038/264045a0

F. Hoyle, N.C. Wickramasinghe, Identification of the λ2, 200Å interstellar absorption feature. Nature **270**, 323–324 (1977). https://doi.org/10.1038/270323a0

F. Hoyle, N.C. Wickramasinghe, Calculations of infrared fuxes from galactic sources for a polysaccharide grain model. Astrophys. Space Sci. **53**(2), 489–505 (1978a). https://doi.org/10.1007/BF00645036

F. Hoyle, N.C. Wickramasinghe, *Lifecloud: The Origin of Life in the Universe* (Dent, London, 1978b)

F. Hoyle, N.C. Wickramasinghe, *Diseases from Space* (Dent, London, 1979)

F. Hoyle, N.C. Wickramasinghe, Comets – a vehicle for panspermia, in *Comets and the Origin of Life*, ed. by C. Ponnamperuma (Reidel, Dordrecht, 1982a), pp. 227–240. https://doi.org/10.1007/978-94-009-8528-5_15

F. Hoyle, N.C. Wickramasinghe, *Evolution from Space: A Theory of Cosmic Creationism* (Simon & Schuster, New York, 1982b)

F. Hoyle, N.C. Wickramasinghe, *Living Comets* (University College Cardiff Press, Cardiff, 1985)

F. Hoyle, N.C. Wickramasinghe, The case for life as a cosmic phenomenon. Nature **322**, 509–511 (1986). https://doi.org/10.1038/322509a0

F. Hoyle, N.C. Wickramasinghe (eds.), *Astronomical Origins of Life: Steps Towards Panspermia*. (Kluwer Academic Publishers, Dordrecht, 2000)

R.K. Kopparapu, A revised estimate of the occurrence rate of terrestrial planets in the habitable zones around Kepler M-dwarfs. Astrophys. J. Lett. **767**(1), article L8 (2013). https://doi.org/10.1088/2041-8205/767/1/L8

T.S. Kuhn, *The Structure of Scientific Revolutions*, 3rd edn. (The University of Chicago Press, Chicago, 1996)

M. Mayor, D. Queloz, A jupiter-mass companion to a solar-type star. Nature **378**, 355–359 (1995). https://doi.org/10.1038/378355a0

H.D. Pflug, Ultrafine structure of organic matter in meteorites, in *Fundamental Studies and the Future of Science*, ed. by N.C. Wickramasinghe, foreword by Fred Hoyle (University College Press, Cardiff, 1984), pp. 24–37

M.A. Rappenglück, „Astronomische Ikonographie" im Jüngeren Paläolithikum (35.000–9.000 BP). Acta Praehist. A. **40**, 179–203 (2020). https://doi.org/10.11588/apa.2008.0.71507

I. Reche, G. D'Orta, N. Mladenov et al., Deposition rates of viruses and bacteria above the atmospheric boundary layer. ISME J. **12**, 1154–1162 (2018). https://doi.org/10.1038/s41396-017-0042-4

P. Slijepcevic, N.C. Wickramasinghe, Reconfiguring SETI in the microbial context: panspermia as a solution to Fermi's paradox. *BioSystems* **206**, article 104441 (2021). https://doi.org/10.1016/j.biosystems.2021.104441

E.J. Steele, S. Al-Mufti, K.A. Augustyn et al., Cause of Cambrian explosion – terrestrial or cosmic? Prog. Biophys. Mol. Biol. **136**, 3–23 (2018). https://doi.org/10.1016/j.pbiomolbio.2018.03.004

E.J. Steele, R.M. Gorczynski, R.A. Lindley et al., Lamarck and panspermia – on the efficient spread of living systems throughout the cosmos. Prog. Biophys. Mol. Biol. **149**, 10–32 (2019). https://doi.org/10.1016/j.pbiomolbio.2019.08.010

E.J. Steele, R.M. Gorczynski, R.A. Lindley et al., Cometary origin of COVID-19. Infect. Dis. Ther. **2**(1), 1–4 (2021). https://doi.org/10.31038/IDT.2021212

R. Temple, The prehistory of panspermia: astrophysical or metaphysical? Int. J. Astrobiol. **6**(2), 169–180 (2007). https://doi.org/10.1017/S1473550407003692

J. Wallis, N.C. Wickramasinghe, D.H. Wallis et al., Physical, chemical and mineral properties of the Polonnaruwa stones, in *Proceedings SPIE*, vol. 8865. Instruments, Methods, and Missions for Astrobiology, vol. XVI (2013), pp. 1–24. https://doi.org/10.1117/12.2029607.

M.K. Wallis, N.C. Wickramasinghe, Interstellar transfer of planetary microbiota. Mon. Not. R. Astron. Soc. **348**(1), 52–61 (2004). https://doi.org/10.1111/J.1365-2966.2004.07355.X

L. Weinstein, Influenza — 1918, a revisit? N. Engl. J. Med. **294**, 1058–1060 (1976). https://doi.org/10.1056/NEJM197605062941911

N.C. Wickramasinghe, The astrobiological case for our cosmic ancestry. Int. J. Astrobiol. **9**(2), 119–129 (2010). https://doi.org/10.1017/S1473550409990413

N.C. Wickramasinghe (ed.), *Vindication of Cosmic Biology: Tribute to Sir Fred Hoyle (1915–2001)*. (World Scientific Press, New Jersey, 2015)

N.C. Wickramasinghe, Giant comet C/2014 UN271 (Bernadinelli-Bernstein) provides new evidence for cometary panspermia. Int. J. Astron. Astrophys. **12**, 1–6 (2022a). https://doi.org/10.4236/ijaa.2022.121001

N.C. Wickramasinghe, Panspermia versus abiogenesis: a clash of cultures. J. Sci. Explor. Anomalistics Front. Sci. **36**(1), 121–129 (2022b). https://doi.org/10.31275/20222199

N.C. Wickramasinghe, F. Hoyle, D. Lloyd, Eruption of comet Hale-Bopp at 6.5 AU. Astrophys. Space Sci. **240**, 161–165 (1996). https://doi.org/10.1007/BF00640204

N.C. Wickramasinghe, E.J. Steele, D.H. Wallis et al., Footprints of past pandemics in the human genome. Virol. Curr. Res. **5**(4), 1–5 (2021)
N.C. Wickramasinghe, J. Wallis, D.H. Wallis, A. Samaranayake, Fossil diatoms in a new carbonaceous meteorite, arXiv:1303.2398v1 [q-bio.OT], https://doi.org/10.48550/arXiv.1303.2398 (2013)
D.K. Yeomans, *A Chronological History of Observation, Science, Myth, and Folklore*, World Science edn. (Wiley, New York, 1991)

Nalin Chandra Wickramasinghe (UK) is a world-renowned Sri Lankan-born British mathematician, astronomer, and astrobiologist. His career spans more than six decades. He was a fellow of Jesus College, Cambridge and a founding member of the Institute of Astronomy in Cambridge, a professor and Head of Department at University College Cardiff and the University of Cardiff, a visiting professor in Sri Lanka and Japan, and he currently holds a honorary position at the University of Buckingham. He has advanced the fields of astronomy and astrobiology and has received numerous awards.

The Cosmic Life Instinct Points the Way to a Healthy Ecological Civilization

Attila Grandpierre

Abstract Ervin Bauer formulated the most promising version of general theoretical biology in a mathematical form. He derived all the basic biological equations from a single fundamental principle, which is known as the Bauer principle. As this principle tells us, living organisms actively and continuously mobilize their free energy content to maximize their distance from lethal thermodynamic equilibrium. Although, for historical reasons, very few people know this principle, and even fewer recognize its power that originates from the fact that it transcends the framework of physics, this universal life principle represents a significant advance in our understanding of the structure and nature of the Universe, even more than the Copernican turn. This comprehensively life-centered worldview unites our physical, emotional, and intellectual aspects. It is also comprehensive in unifying individual, communal, and cosmic life. It offers an unexpectedly profound scientific basis for a cosmic ecology respecting all life forms, including the Living Universe. It offers new perspectives for our conduct of individual life as well as for good government and developing a healthy civilization.

Keywords Scientific method and explanation · First principles · Principle of least action · Principle of greatest action · Principle of life · Worldview shift · Healthy civilization

---

A. Grandpierre (✉)
Budapest Centre for Long-Term Sustainability (BC4LS), Budapest, Hungary
e-mail: attila.grandpierre@bc4ls.com

D. Bartosch et al. (eds.), *Towards a Philosophy of Cosmic Life*,
https://doi.org/10.1007/978-981-99-2131-7_4

Introduction and Brief Overview

> There is one natural law for all of us to live by, and that is caring and sharing and
>
> being responsible for the wellbeing of all other living things…
>
> Having a good relationship with all life is the key to
>
> physical, mental, social, and spiritual harmony.
>
> This is the way it was meant to be.
>
> Tjilpi Bob Randall, Anangu leader

The most effective elements of science are the *most general laws* of Nature. The discovery of the *physical laws* was an event of unique importance in the history of humankind which set the modern scientific revolution in motion and enabled Western civilization to dominate the world. I will present a series of arguments to consider, indicating that from time immemorial, human cognition has necessarily been directed toward the exploration of natural laws and the even deeper explanatory *principles* behind them. This is extremely important for us because the last century has seen the emergence of new disciplines investigating the relationship between life and the Universe, the consensus of which has led to the realization that we live in a Living Universe. In this context, I introduce here an unexpectedly profound theoretical biology that arrived at a mathematical formulation of a *universal biological principle* and the derivation of all the biological laws of Nature from it (Bauer 1967). Remarkably, this biological principle explains the convergence of results in these new disciplines (Grandpierre 2021a, b). The multifaceted agreement scientifically refutes the basic assumption of the physicalist worldview, making the qualitative transformation of the scientific worldview inevitable. Changing the dominant worldview transforms the type of civilization entirely. Ervin Bauer's theoretical biology is the foundation for a scientific and historical turn that is more profound than the Copernican one. While the first scientific revolution turned civilization toward a matter-principled, substance-based, individualistic way of thinking, the second, biological revolution will turn history toward a life-principled, principle-based, community-affirming one.

The Aim of Universal Human Cognition is to Understand Nature as a Whole on the Basis of Explanatory Principles

Strangely, in recent centuries the extreme importance of the fundamental role of first principles in science became obscured; and we lost the understanding that such principles were known before the first civilizations known today.

Furthermore, it is not sufficiently well acknowledged that in ancient Mesopotamia, even before the Greeks, a formulation of the three principles of the Universe and their unity, the cosmic Trinity, was recognized. "[T]he first unique principle from which all the other gods took their origin, was [...] the One and Good [...] the first person of the supreme trinity [...] the primordial principle" (Lenormant 1999, 113–114; referring to texts on cuneiform tablets from Ashurbanipal's library at the royal palace at Nineveh from before 625 BCE; please note that Thales, the so-called first philosopher, was born around 620 BCE). Neither Thales nor the later Greek philosophers knew what these principles were, but they knew that to know them was to know *the essence of the Universe*, and therefore their activity was directed to the discovery of these *arkhai* (ἀρχαί).

The Role of the Soul, the Intellect, and the Life Instinct in Our Decisions

There are timeless facts of knowledge. It is such a timeless fact that life requires action, action requires decision, and adequate decision requires reason. *The property of life that drives us, from within, to act, can be called the life instinct. The aim appropriate for life is provided by the life instinct, the possibility of realizing that aim is provided by the world. If the world is that 'what is,' then it is the life instinct that tells us 'what should be.' The basic task of all living beings is to make choices appropriate to life and the world. The life instinct is conceived by the soul, the world by the mind.* To be able to interpret and evaluate the feelings conceived by the soul and the data conceived by the senses in accordance with the life instinct, man needs a uniform basis of reference. Naturally, this reference basis is the life instinct. For a realistic perception of the life instinct, a pure soul is necessary. A realistic perception of the external world, adequate for the aims provided by the life instinct, requires a pure intellect.

The human intellect is able to take the mentally created image of the life instinct and the mentally created image of the world as a reference point instead of the life instinct and the world themselves. If neither the soul nor the intellect provides a true image of the life instinct and the world, then our conscious decisions and actions do not correspond to Reality.

The Central Role of the Most General Laws of Nature in Understanding Reality

For man to be both a living being and an intelligent being, we need a thorough understanding of life and Nature that is comprehensible and applicable to our everyday lives. The basic task of the human intellect is to understand the interconnections in life and the world. Since time immemorial, man's task in life has been to perceive and faithfully reflect the predictable changes in the external and internal environment. Changes in natural phenomena are governed by the laws of Nature. Natural laws permeate everything that exists, are everywhere and at all times, and fundamentally determine the course of phenomena. To shape life in a way that is in keeping with the instinct of life, it is necessary to be aware of these rule-governed changes, because they help us predict the course of phenomena and thus shape our actions in a conscious and purposeful way. Ancient man, by his very nature, was deeply dependent on the knowledge of the laws of Nature, and thus strove to recognize, understand, and to explain these laws.

The Explanatory Chain of Nature is Based on Fundamental Principles

It is also a timeless fact that it is in the nature of explanation to explain the much by the little. The better an explanation, the more it explains with the less. *The best explanatory system is the one that also explains the natural laws of observable phenomena, and does so with a single, most general, and deepest natural law. These deepest laws of Nature, which together can provide a single, comprehensive explanation of the whole of Nature and all its fundamental aspects, are called first or fundamental principles.* It is only by understanding all the first principles together and how these principles form a unified whole that we come to understand Nature as a whole. Therein lies the basic task of human life. Incredible as it may seem in today's world, understanding the explanatory chain of Nature as a whole has always been, as it is today, the most fundamental task of human life.

Let's illustrate this with an example. The Hungarian biologist, ethologist, and evolutionary theorist Vilmos Csányi pointed out that the basic function of the nervous system is to coordinate the connections between perception and motor functions and to make the decisions necessary for life. The neural network maps and models changes in the external environment. The operation of the model enables the animal to anticipate predictable changes in the environment that are important to it and to make behavioral decisions based on these anticipations (Csányi 1994, 134). More closely, the basic task of the nervous system is to make 'good' decisions, decisions that are necessary for a life that are in satisfactory harmony with the life instinct. 'Good' decisions are born from 'good' feelings. A decision and feeling are 'good' when they correspond to the life instinct, and the better they correspond, the better the decision is and the better it feels.

In Ancient Nature Religion, the Natural Order is Based on the Life Principle

Naturally, the instinct for life played a central role in the worldview of the ancient peoples. In ancient societies, religion of nature also professed that the whole of Nature and every part of it is endowed with soul. Because the soul had been essentially the same as the instinct for life and the feelings that come from it, the oldest religion of humankind had been based on the realization that the instinct for life permeates all of Nature. Since the 19th century, this ancient and cosmic nature religion has also been referred to as the religion of the soul, and in the jargon of experts it is called animism. Beginning in ancient Eurasia, and in many other parts of the world, animism became a scientific theory in which the soul was identified with the life principle, which was conceived as the source of all fundamental laws of Nature and society. "In the traditional thinking of so-called archaic societies, an immanent power, a vital principle, an individualized dynamism, is usually recognized to exist not only in man but in certain other material and biological beings as well" (Riviere 1987, 426). In fact, even today there are indigenous peoples who base their worldview on the life principle and identify it with the life force (Patterson 1998). This ancient religion of nature stems from an attitude that is extremely sensitive to the mystery of life energy hidden in all beings. The ancient Japanese nature religion Shinto turns with the highest reverence toward this mysterious energy that also awakens astonishment and admiration. Shinto refers to the law of the natural order that permeates all of Nature, and if one understands this, one can learn from it the divine and human nature of life and also how to live our lives. The moral virtues of the natural order, according to Shinto, are honesty and purity, writes Bhikkhu Satori Bhante in his book about Shinto. And he adds that the guarantee of happiness is the purity of the heart (Bhante 1990). Since the instinct for life is conceived by the soul, a clear perception of the life instinct requires a pure soul, a pure heart.

The ancient aspiration of humankind to recognize the deepest law of Nature is also indicated by the fact that many indigenous peoples regarded the cosmic law as the basic law of society (Long 1987).[1] Indeed, they are either free to choose the laws of society or to conform to nature-given Reality. If social laws have no universal basis in Reality, then there is no reason to obey them. Moreover, the fact that humans are potentially living and intelligent beings is a natural fact; since society is based upon humans, it should be based on this fact. Humankind, as natural beings, is under the control of the laws operating in the causal system of Nature. And if we want to bring society into harmony with the nature of humankind, it must conform to the universal laws of Nature. Ancient peoples shaped their lives, societies, and ways of life in the spirit of this recognition. Sturm holds the view that, in its broadest sense, the law of Nature is the rule that governs action and is derived from the essential properties of Reality. From ancient cultures to the present day, man has turned again and again to the law of Nature for measure, direction, purpose. Knowledge of natural law enables us to be true to our true being (1987, 318). In his book *The Origin of Things*, Julius Lips writes:

> The world of primitive peoples is based on the belief: "In the beginning was the power." This wonderful power is omnipresent, and its existence is as certain as the hardness of stone and the wetness of water, as all-pervading as the ether in modern physics. This power, supernatural only to modern man, but completely real and natural to the primitives, is known to them by many different names. To the Malayo-Polynesians it is mana, among the Iroquois Indians it is orenda, among the Sioux, wakan, among the Algonquians, manitou. To recognize the

workings of this power [the vital power], to have part in it, to use it, and to master it—these are the aims of primitive man. In his world there are no coincidences—everything has its causes and associations; to discover these is the task of man. In the minds of primitive men, cause and effect are not restricted to the small domain of the physical world [in contrast to the modern Western mindset], but they are associated with the powers and phenomena beyond the visible world. There is a constant tendency among the natural peoples to recognize the manifestations of this [invisible] force. (Lips 1962, 312, tr. and insertions in brackets AG)

It is noteworthy that indigenous communities, such as the Maori people of New Zealand, knew the life principle and based their long-term sustainable society on it, developing their environmental philosophy upon their philosophy of life (Patterson 1998). According to the Maori,

we should respect the natural world because each creature has a mauri or life force, and that this life force joins all beings—humans, gods, plants and animals, mountains and rivers and seas—into one interdependent whole, each part depending for its well-being upon the health of each other part and of the whole itself. (Patterson 1998, 70)

It seems more evident to us as the century draws to a close that people from the traditional cultures of the East, of Africa, of North-America and of Australia have developed and preserved attitudes towards the natural world from which we have much to learn. What is common to them is a feeling for the affinity between the human and the natural world, and a sense that all things belong together and work together like the organs of a living thing. Animism, the belief that nature is activated by spiritual forces, is ancient and widespread, and can be seen in the sacred writings of the Hindus as well as in the oral traditions of the North American Indians. (Clarke 1993, 17)

All this suggests that the constant endeavor of primitive peoples has been to recognize the causal system of the invisible life force in addition to that of the material forces. There are indications that ancient humankind succeeded in this endeavor and developed a broader vision and a deeper explanatory world-view than the materialistic view that prevails today. Indeed, in the modern world, we know more and more about the details and less and less about the most important thing to know, that is, about the whole of life and the world. The current prevailing view in Western philosophy is expressed by "the doctrine [which] tells [us] that all there is to the world is a vast mosaic of local matters of particular fact, just one little thing and then another" (Lewis 1986, ix, insertions in brackets AG). Comparing this picture with that of indigenous peoples, we can conclude that the scope of their worldview exceeds that of the modern one.[2]

The Main Path of Science and Philosophy Has Been Directed, for Millennia, to Discover the Fundamental Principles of the Universe

The Highest Goal of Science and Philosophy is to Know the Whole of Nature and to Explain it by Fundamental Principles

It is also a timeless fact that there are two "directions" to be taken in exploring the explanatory system of the world. First, one can start from observable phenomena and then investigate further to find the laws of Nature, and deeper still, until all the fundamental principles are found. This opens the way to the second direction, which is to apply these principles and laws to concrete cases.

A solid foundation for our explanatory system is vital, because it determines how good our lives will be. We need a coherent system of the best-established, most reliable knowledge. The importance of such a reliable system of knowledge is so paramount that it has been given a special name: science. Science relies on both types of human knowledge, empirical and theoretical, for maximum reliability. The basic human endeavor to understand Reality as a whole and as clearly as possible is therefore closely linked to scientific thinking. Surprising as it may seem, history seems to broadly support this conclusion.

The awareness of the extraordinary importance of first principles has determined the main road of philosophy and science for more than two millennia after Mesopotamian and pre-Socratic philosophy—and, as I argue above, before it as well.

> The basic idea is [that] there are two "directions" to proceed in our methods of inquiry: one away from what is observed, to the more fundamental, general, and encompassing principles; the other [leads] from the fundamental and general to [possible specific instantiations of those] principles. *The basic aim and method of inquiry identified here can be seen as a theme running throughout the next two millennia of reflection on the correct way to seek after knowledge: carefully observe nature and then seek rules or principles which explain or predict its operation.* (Hepburn and Andersen 2021, italics and insertions in brackets AG)

The search for Nature's first principles has been a common thread in the history of science for at least 2600 years, with mathematics and physics proving to be the most successful. This method was known in some form by Plato and Aristotle, and, among many, reformulated by Francis Bacon and Descartes, the fathers of modern scientific method. It is this ancient insight that was put into the following words by the American philosopher of science Filmer Stuart Cuckow Northrop (1893–1992) in the early 20th century:

> Science proceeds in two opposite directions […]. It moves forward with the aid of exact mathematical formulation [of natural laws] to new applications, and backward [from observable phenomena and natural laws] with the aid of careful logical analysis to first principles. *The fruit of the first movement is applied science, that of the second theoretical science. When this movement toward theoretical science is carried through for all branches of science, we*

> *come to first principles and have philosophy.* (Northrop 1931, 1; italics and insertions in brackets AG)

Similarly, George Gaylord Simpson (1902–1984) wrote in his paper published in the prestigious journal *Science*:

> *Bacon further maintained that the unity of nature would be demonstrated and the sciences would be incorporated into one general body by a fundamental doctrine, a Prima Philosophia, uniting what is common to all the sciences* [...]. In our own days, Einstein and others have sought unification of scientific concepts in the form of principles of increasing generality. The goal is a connected body of theory that might ultimately be completely general in the sense of applying to all material phenomena (Simpson 1963, 87).

The progress of science, in its effort to discover the laws of Nature and then the principles underlying these laws of Nature requires induction, the inference of a general law from particular instances. The inductive method leads to theoretical science. Once theoretical science has uncovered the general laws of Nature, and, in the best case, even the fundamental or first principles, the next step is to apply the laws/principles obtained by theoretical science, and to 'deduce' the conclusions, for example, predict specific phenomena on the basis of the general law/principle. This step is called deduction. *Inductive reasoning leads to theoretical sciences and to an understanding of the nature of Reality.* Deductive reasoning works from the most general laws and principles and leads to applied sciences. When the perfection of theoretical science is carried out in all the fundamental branches of science, and we come to the discovery of the unity of all the fundamental principles of Nature, philosophy, in the form of the unified science capable of understanding all the most general aspects of the Universe, falls into our hands. Through such an all-comprehensive science, which is also philosophy at the same time, the nature of ultimate Reality and the ontological architecture of Reality in its entirety becomes clear. Without any further step, this all-comprehensive science yields an all-comprehensive worldview, an adequate, true picture of the Universe, meant as the unified whole of everything that exists. It is on this rock-solid knowledge of the ultimate Reality that we should build our conduct of life; it is this all-comprehensive science, philosophy, and worldview that is capable of yielding the directive for good governance of societies; it is the all-comprehensive worldview capable of establishing a healthy type of decision-making systems and a long-term sustainable, healthy civilization.

Definition of Health

By health we mean a high level of physical, emotional, and intellectual well-being, including our physical, emotional, and intellectual capacity to cope with our life situations and to develop in the whole system of our relationships, including the high quality of well-being in our relation to our cells, ourselves, our family, our nation, humanity, the living world on Earth and the Living Universe. We cannot buy this good relationship in the shop. This personal relationship is partly shaped by our genetics, family, and social influences, and partly by ourselves. It is in our most personal, elementary interest to perceive life and the world in the highest possible quality and to give the highest possible quality response to both life and the world.

Achieving and maintaining good health requires continuous efforts that challenge our best abilities. We can only develop through persistent, conscious efforts, from which we can learn how to do better. Our physical well-being is our high quality and broad range of our capability to act. Our emotional well-being is our deep happiness. Our intellectual well-being is our clarity of vision, depth of insights, and the high level of our capacity of solving problems. "The aim and meaning of life are to live usefully in ceaseless development and constant perfection" (Kártyikné 1989, 6, tr. AG). "We are on the right path when we form our individual interests into a common goal" (11, tr. AG).

Sustainability requires a motivation that is at least as strong as the attraction of owning material things (Budapest Centre for Long-term Sustainability 2021). The fact that the life principle is more fundamental than the material principle provides a scientific basis for a motivational system that is stronger and richer than the motivational system for material things. The quality of health is fundamentally characterized by the depth of our feeling and thinking. The deeper we feel and think, the stronger our capacity for motivation. The more broadly we think, the more many-sided and the richer our motivational system becomes. We need a worldview that allows for the deepest and most multi-faceted emotional and intellectual world, the strongest and richest motivational system, and thus enabling us to fulfill our creative potential and achieve the health of our capacity to cope and develop.

Characteristics of a Healthy World View

(1) A worldview is healthy if it is complete at the ontological level, that is, if it corresponds to the whole of life and the Universe as a whole, gives a correct picture of all three fundamental aspects of the Universe, is thorough in terms of its explanatory system down to the fundamental principles, and is balanced. Balanced means that it recognizes the primacy and fundamental nature of the life principle as well as the essential nature of the material principle and the principle of reason. It is able to provide the solid ground, the driving force, the energy, and the compass needed to ensure health in all life situations because it is rooted in the immutable and universal reality of the life principle; and the life principle is capable of realizing its ultimate purpose, the fulfillment of life.

(2) The life principle, which underpins a healthy worldview, encourages us to continuously improve our overall quality of life, to achieve the fullness of our health, and to mobilize all our life energies and capacities to this end.

(3) Since the life instinct has a cosmic scope, it connects us most intimately to each other, to all living things and to the Universe, and inspires us to uplift all life, to recognize and realize the greatest potential of universal life and our personal lives.

Let us keep in mind that the Universe, that is, Nature, is the unified whole of all that exists, or "the" Reality. If we can arrive at a science of Nature in which all of the fundamental disciplines have an explanatory system based on the same, single fundamental principle, and these fundamental disciplines together form a unified, all-comprehensive whole, then, as for example Francis Bacon, Filmer Stuart Cuckow Northrop, and George Gaylord Simpson have suggested, we arrive at a scientific understanding of Reality, a scientific foundation for philosophy, the *prima philosophia*, the *summum bonum*, that is, at an explanation of Reality as a whole.

The Explanatory System of the Natural Sciences is Based on Fundamental Principles

The Principle of Least Action is the Fundamental Principle of Physics

In the 20th century, it became clear that elementary particles are not really elementary, namely because they are based on universal force fields of interactions (Einstein and Infeld 1938; Weinberg 1995). According to quantum field theory, elementary particles are created by universal force fields of interactions described by the fundamental equations of physics. These equations can be derived from the principle of least action (as discussed by Richard P. Feynman in his famous doctoral dissertation in 1942, see also Coopersmith [2017]). The most fundamental quantity of the action principle, the action, is a "cost function" (Rosen, 1967, 4). This special cost function measures the product of energy investment and time investment. Remarkably, these two quantities are of vital importance for all living organisms; these are the two most important factors for the performance of an action. The first formulator of the principle of least action in the history of modern science, Pierre Louis Moreau de Maupertuis (1698–1759), ascribed to the 'action' in the principle of least action the following meaning: "the unity of vital force in the Universe" (Jourdain 1912, 426).

The principle of least action is well known in physics, but the fact that *all* the fundamental equations of physics can be derived from it is rarely recognized. And the outstanding fact that it is precisely the defining property of the fundamental principle that can be used to derive all the fundamental equations of a given discipline seems to have been overlooked in the history of modern physics. The situation can be described by the following quotations:

> The action principle of physics is claimed to be, regarding its form and content, to come nearest to the ideal final aim of theoretical research to condense all natural phenomena into one simple principle, that allows the computation of past and future processes. (Yourgrau and Mandelstam 1955, 126)

> We have found that the Principle of Least Action applies across all scales, from the realm of the microscopic (actually, tinier than that) to the everyday (classical mechanics, engineering, optics, the transmission of radiation, physical chemistry, statistical mechanics, continuum

> mechanics), and on to the whole cosmos (gravitation due to stars, planets, black holes, and gravity waves). (Simpson 1963, 195)

> The action principle turns out to be universally applicable in physics. All physical theories can be formulated in terms of an action. Moreover, the formulation of action is elegantly concise. The reader must understand that the whole physical world is described by a single action. (Zee 1986, 109)

It is indisputable that "the principle of the least action leads to the whole of physics" (Coopersmith 2017, vii).

Despite all of these arguments, the principle of least action currently receives little attention among philosophers (Stöltzner 2003). Apparently, this absurd situation is rooted in the economic and consumerist interests that dominate materialistic societies. The principle of least action is still considered a "surprisingly well-kept secret," even though "these ideas are at the heart of the physical sciences and engineering" (Coopersmith 2017, 1). In fact, it is not only that the first principles of science, despite the considerations indicated above, receive undeservedly little attention. The mainstream of modern science and philosophy in recent centuries has explicitly taken the position of rejecting knowledge that allows us to understand the foundations of Reality, ignoring the importance of first principles. Actually, as the Scottish-American philosopher Alasdair C. MacIntyre, Emeritus Professor of Philosophy at the University of Notre Dame, characterizes the present situation of philosophy, "*Nothing is more generally unacceptable in recent philosophy than any conception of a first principle*" (MacIntyre 1990, 4, italics AG).

Given that philosophy studies the most general aspects of Reality, and that the principle of least action is the only universal principle in physics, it seems almost inevitable that philosophy should recognize the all-encompassing importance of the principle of least action. In this work, this very important step is taken: *I take the principle of the least action as the fundamental principle of physics.*

The Present State of the Explanatory System of Physics

Paul Davies, the well-known physicist and cosmologist, describes the state of the explanatory system of physics which still applies today: *Almost all physicists who deal with fundamental problems seem to accept that the laws of physics are "at the base of the rational explanatory chain, in the same way that the axioms of Euclid stand at the base of the logical scheme we call geometry*" (Davies 2004, 5, italics AG). This means that the explanatory chain of modern science stops at natural laws. Let us note that axioms are qualitatively different from laws. For the system of axioms, in mathematics and logic, is the source of laws. The system of axioms plays the same role as the principles. Thus, for example, from the axioms of Euclid it is possible to derive all the laws of (Euclidean) geometry, just as from the axioms of logic all the laws of logic. In the explanatory chain, axioms are found at a deeper level than laws.

By stopping the explanatory scheme of physics at the level of laws, modern physics has taken a step backwards along the millennia-long main road of scientific progress.

This regress may be responsible for the lack of new milestones in the century that has passed since the discovery of quantum physics. Actually, the main road of science is characterised by the discovery of deeper and deeper laws of Nature. This is the road that led to the discovery of biology's fundamental principle.

The Fundamental Principle of Biology, the Principle of Life

While the evolution of physics has stalled at the level of physical laws, thanks to Ervin Bauer, a Hungarian-born biologist, biology reached the mathematical formulation of its fundamental principle. Bauer showed that all the basic biological laws can be derived from a single unifying, universal, and overarching principle. This biological principle was later called the Bauer principle. The Bauer principle states that "*[l]iving, and only living, systems are never in equilibrium, and, at the expense of their free energy content, are constantly working against the equilibrium demanded by the laws of physics and chemistry under the given external conditions*" (Bauer 1967, 51, tr. and italics AG). *This means that biology has its own principle, which cannot be derived from the principle of physics, since the essence of life is found precisely in systematic deviations from physical behavior.*

By supplementing Bauer's work with the most powerful tools of modern physics, it turned out that Bauer's principle itself could be derived from a more general principle, the principle of greatest action that involves also the mathematical formulation of biological autonomy (Grandpierre 2007, 2012, 2013; Grandpierre and Kafatos 2012; Grandpierre et al. 2014). The fundamental principle of physics is only a special limiting case of this biological principle, namely the case in which the capability to act is below the Planck limit of quantum-level indeterminacy. This is the reason why the behavior of macroscopic physical bodies can be characterized as inert, or passive. And since the principle of greatest action pervades the Universe as much as the principle of least action, it can be considered a cosmic principle. The two principles prevail simultaneously, since the principle of greatest action works at a more fundamental level—beyond the quantum vacuum—by biologically organizing the spontaneous production of virtual particles which are the sources of physical interactions governed by the principle of least action (Grandpierre 2007, 2014). In other words: the principle of greatest action is the principle governing the biological processes selecting biological aims, while the principle of least action is the principle governing the physical processes that execute these aims.

If we further generalize the principle of greatest action to include the truly biological state variables, that is, the feelings and thoughts that give rise to biological motives, we obtain the life principle. It is the principle of biology and has an all-comprehensive, unifying, and cosmic scope. The biological unification effect is due to the fact that life is manifested in "units," that is, living beings; the unification is the result of the co-*operation* of biological autonomy and the life principle, possible under the right conditions, leading to biological organization according to biological aims. This cosmic scope of the life principle implies that the Universe is essentially not inanimate but living. Rather than being "lazy" (Coopersmith 2017), the Universe is continuously active beyond the quantum level of Reality, manifesting its living characteristics. Biological action is made possible by the organization in living organisms, which puts the spontaneous production of quantum-level virtual particle pairs at the service of biological aims (Grandpierre 2007, 2012, 2021a, b). Cosmic life "starts" in the microcosm; this is where the diversity of cosmic life is at its greatest.

At the time when Copernicus's work was finally accepted, this was the result of a convergence of new scientific discoveries that had reached a critical mass: the more precise measurements of Tycho Brahe, the arguments of Giordano Bruno, Galileo's experiments, the decisive achievements of Francis Bacon and René Descartes in the philosophy of science, Kepler's laws and then Newton's laws, all in harmony with each other, confirming each other. These agreements had a tremendous influence on the scientific method; they shifted the main aspects and methods of scientific research towards a new, physicalist basis. The situation is now similar to that of the Copernican turn. Independently of each other, a whole range of new disciplines and fields of research, each with their own hitherto unimaginable scope, have emerged on the scene to investigate the relationship between life and the Universe. These are the following: (1) the biocentric cosmology of the eminent biochemist Joseph Lawrence Henderson (Henderson 1913; Grandpierre 2021b, 239–253); (2) anthropic cosmology, which is somewhat misnamed since it is not specified to humans but actually investigates the cosmic role of *life in general* by means of cosmological models (Barrow and Tipler 1986; Grandpierre 2021b, 289–301); (3) laboratory experiments on the origin of life, empirical research showing the quick origin of life on the young Earth, and Oparin's conclusions regarding the existence of an autonomous law of life (Oparin 1960, 5–12; Grandpierre 2021b, 253–283); (4) the theory of a bio-friendly universe (Davies 1998, 2003, 2006); (5) *astrobiology,* the science that studies the relationships between life and the universe, arriving at the general view of astrobiologists that "life is a cosmic imperative" (Davies 1998; Grandpierre 2021b, 284–289); (6) the Gaia theory, according to which Earth is a living system and self-regulating within a range favorable for life (Lovelock 1987; Grandpierre 2021b, 301–322); (7) recent advances in biology (e.g., Tompa and Rose 2011; Grandpierre 2021b, 309); (8) Ervin Bauer's theoretical biology (Bauer 1967 [first 1935]), complemented by the generalization of the least action principle, with an exact mathematical formulation of teleology and (9) the formulation of the principle of greatest action (Grandpierre 2007), and its further development with the quantitative theory of biological, autonomous decision-making (Grandpierre 2012, 2013, 2021b, 169–237; Grandpierre and Kafatos 2012; Grandpierre et al. 2013, 2014). The convergence of these new approaches provides solid scientific evidence, from both theoretical, observational, and experimental perspectives, that *there is a law of Nature that governs the development of the conditions toward becoming favorable to life* (Grandpierre 2021b, 243–322).

These new disciplines thus confirm the fundamental role of the life principle that ensures the leading role of life throughout the Universe. Remarkably, the principle of greatest action *explains* the existence of this indirectly detected natural law. The combined weight of these disciplines, together with the theoretical biology founded by Ervin Bauer, is capable of triggering the recognition of the fundamental importance of the life-principled outlook. Based on this, it is difficult to avoid the conclusion that we need to permanently and irrevocably change our perception of the foundations of Reality. This new series of scientific contexts of Bauer's masterpiece opens unexpectedly profound and uplifting perspectives for the future of science, philosophy, and civilization, initiating an epoch-making turn comparable to the Copernican revolution. This is all the more so because a comprehensive science based on the principle of matter, the principle of life, and the principle of reason can offer a cosmically grounded ecology on an exact scientific basis and thus scientifically underpin the ecological turn (Budapest Centre for Long-term Sustainability 2021).

Explaining the Independent Decision-Making of Living Organisms

Most physicists believe that the law of conservation of energy is violated within the limits of indeterminacy given by Heisenberg's uncertainty relation (Davies 1984, 104–105). In addition, it is a general view that virtual particles arise spontaneously, without *any* cause. Accordingly, this process is called 'acausal.' However, the principle of causality knows no exceptions. And if there really is no physical cause for the production of virtual particles, there must be another natural cause. In natural science, the next obvious possibility is that, within living organisms, biological causes are responsible for the spontaneous production of such virtual particles. This is all the more so since in biology spontaneous phenomena are a fundamental feature of biologically motivated processes, indicated also by Bauer's theoretical biology. And since, in accordance with the life principle, living beings have biologically controllable energy, such as the so-called volitional energy, they are capable of providing the energy necessary for the creation of virtual particles.

With the life principle and biological autonomy, we have, without any special effort, an explanation that, in one stroke, remedies the two fundamental shortcomings of quantum physics: the violation of the universal principles of conservation of energy and causality. At the same time, and even more importantly, it can explain the physics of biological causality, that is, teleology. It is able to explain the physical realization of an action that can be triggered by biological and intellectual causes. Biological autonomy, through biologically controllable energy behind our ability to make decisions, makes living beings capable of creating virtual particles through decisions corresponding to biologically controllable energy. Recent psychological research has confirmed that our will is indeed capable of mobilizing our free, biologically mobilizable energy (Baumeister 2012). It is also remarkable that the principle of least action itself works through virtual particles. Biological autonomy is able to ensure that physical processes take place according to our will by creating virtual particles that are in accordance with our will (Grandpierre 2012; Grandpierre and Kafatos 2012, 2013).

Physics, Biology, and Logic Are All Based on Fundamental Principles

Let us now note that not only the sciences of matter and life, but the science of reason, namely, logic, also manifests an axiomatic structure. Logic is the science of correct reasoning, that is, the science of drawing correct conclusions from previous statements, called premises,[3] on the basis of logical laws. Remarkably, this premise–law relation also exists in physics. Physical laws determine the behavior of physical bodies together with the given external conditions; in numerical calculations, the differential equations of physics work with given input conditions. The input conditions of physical equations can be set in parallel with the premises of the laws of logic. In biology, the situation is similar; in the case of living organisms, the decisions arising from biological autonomy play the role of input conditions.

When the system of the axioms of logic is generalized into a principle of reason that works with premises arising both from "external" as well as "internal" observations of natural rational and emotional motivations arising from the life principle, then it becomes clear that physics, biology, and logic are all based on fundamental principles. All three have an explanatory system with three levels corresponding to phenomena, natural laws, and principles. In the comprehensive framework unifying the sciences of matter, life, and reason, it becomes clear that the Universe is based on the principles of matter, life, and reason. In this way, we have arrived at a comprehensive, ontologically complete, scientific understanding of Nature, so that the scientific basis of philosophy falls into our hands.

The Thrice Threefold, Hierarchical Structure of the Universe

It is important to distinguish between, on the one hand, the equations of theoretical sciences, which "exist" only in our mind and become more and more precise with the development of science, and, on the other hand, the natural laws they describe, which are their equivalents existing in Nature. Let us now consider the relationship between the explanatory chain of science and its equivalent in Nature. Let us add that the equations of natural sciences are laws of changes, and changes are due to causes. The explanatory chain of science is directed toward discovering the causes acting in Nature. The most important cause is the first cause.

In the light of these considerations, it becomes clear that the explanatory chain of science mirrors the causal chain of Nature, but in reverse order. Accordingly, the first causes of Nature are found in the first principles and their input conditions. We have arrived at the causal structure of the Universe. We have found it in a triple trinity. This triple trinity is the trinity of phenomena, laws, and principles in physics, biology, and in the science of logical thinking, the latter of which we may call *noology,* since in ancient Greek philosophy the rational mind was called *nous* (νοῦς).

The Seven Cosmic Communities of Life—Communities Within Communities up to the Living Universe

Cosmic Life can act locally in physical reality only through the physical units of action capable of realizing its biological ends. Recognizing that the Living Universe must be capable of action at all spatial scales, and that its only means is the biological organization of the free energies available (Grandpierre 2018), we can realize that in order to be able to live, the Living Universe must be able to act on all scales. This requirement demands the existence of "life units" with appropriate dimensions. The Universe must develop living organisms at microscopic, macroscopic, and cosmic scales. Given that there is an unlimited amount of free energy available in the quantum vacuum, and that this energy can be organized by a life principle that operates beyond quantum physics, small enough units of life capable of operating on microscopic scales are needed. This branch of cosmic evolution led to the origin of the first cells on Earth (Grandpierre 2013). Once living cells are developed on the basis of the cosmic life principle, the life principle also prepares the ground to guide biological evolution toward more complex life. This allows the evolution of living organisms to develop units of life on a significantly larger macroscopic scale, that is, multicellular beings. The presence of planetary-level free energies allows the emergence of planetary-level life. According to comprehensive science, this branch of cosmic evolution has led to the development of the self-regulating Gaia (Lovelock 1987) and to Helios, the Sun, which is alive and has its own activity that is called solar activity, and which is constantly reorganizing its activity according to the Bauer principle (Grandpierre 2015, 2018).

Now let us take into account the fundamental fact that life is present always in the form of communities. The life of the individual is only possible through the previous existence of the family. The existence of the family is based on the existence of the nation, the existence of the nation on the existence of humanity, the existence of humanity on the existence of life on Earth; and the evolution of life on Earth is a branch of cosmic evolution. From the point of view of humanity, there are seven units of life: cell, individual, family, nation, humanity, Gaia, and the Living Universe. These seven levels of organization can be considered the "ladder" of cosmic life. We belong to all of them simultaneously, and we have a responsibility to all of them. It is this interconnectedness that gives human beings their place in the world and in the Universe.

In summary, there are three levels of Reality: phenomena, laws, and first principles. There are basically three kinds of existence, corresponding to matter, life, and rational mind. We have come to a scientific mapping and explanation of the architecture of the Universe, and found it to manifest a 3 × 3 × 7 structure.

From here, we arrive at the next set of questions: Where does this Living Universe come from? If the Universe is a unified whole encompassing all that exists, then it cannot have originated from anything other than its own pre-existence. But how can it arise from itself? How can this question be clarified?

There is Only One Entity that Can Exist by Its Own Power: Cosmic Life

The ancient Greeks called the *self-existent cosmic Reality* the 'first cause,' 'God,' or 'ultimate substance.' The existence of a self-existent, primordial cosmic Reality means that this Reality, acting as its own cause, has created itself. In short, it must be not only self-existing but also self-creating. Note that self-creation is action. Action assumes the presence of life. Consequently, self-creation assumes life. Now, let us also consider that life is not only necessary but also a sufficient condition for Reality existing by its own power, because the essence of life is precisely that it creates life and sustains life, that is, itself. Life has a twofold nature: It not only exists, but also, and this is its essential nature, creates itself. It is that kind of existence that is directed toward creating itself, directed toward developing and realizing its full potentials, the totality of itself. Cosmic life is the self-creating activity of the whole of Reality.

For example, the material or physical universe cannot come into existence by itself; even the Big Bang assumes a 'starting up,' a so-called, extremely specific, pre-existent 'quantum fluctuation' that initiates the process. This process cannot be acausal, so if it does not have a physical cause, it must have a *biological cause*, and this assumes the *pre-existence of life*.

Consequently, cosmic life is eternal. Cosmic life is itself "the work in process" of the 'first cause' which is active in the eternal functioning of cosmic life. Cosmic life is the 'ultimate substance' postulated by the Greeks. In this respect, cosmic life can be considered as divine life. Actually, cosmic life has more 'divine' aspects. It is noteworthy that there is only one life principle in the Universe—in the same way as there is only one physical principle in Nature. This life principle permeates everything that exists, including us as living beings. This fundamental Reality acts 'within' us; from the aspect of our 'self'—defined as our decision-making center—it is the life instinct. The life principle is both immanent and transcendent, since we have an access to the life instinct 'internally,' and it is transcendent, since it is 'beyond' the observable universe. That is to say, the life instinct of living organisms and of the Living Universe are identical. In the causal order of life, the 'first' cause acts at the cosmic level. In the case of an eternal Being we cannot speak of a 'first' cause in time, only in a logical sense. Since the living Universe is the primordial Being itself, we can in the logical sense regard its first division into the material world, the world of life, and the world of reason as the first step of its self-realization. This 'first division' of cosmic life into three kinds of 'realities' is a step in the logical sense, since in Reality they are not only differentiated but also interrelated and integrated. But the extraordinary, ontological depth of their distinguishability indicates that matter, life, and rational mind are the most fundamental features of the Universe.

Let us make the next step. An understanding of the logical relationship between the principle of matter, the principle of life, and the principle of reason brings us closer to a deeper understanding of ontology.

On the Relationship Between the Principle of Least Action and the Principle of Greatest Action

Let me illustrate the relationship between the principle of least action and the principle of greatest action by means of a simple example.

A company building bridges aims to build as many bridges as possible every year. This corresponds to the principle of greatest action. It can only build as many bridges as possible if it builds each bridge in the most economical way possible. It is precisely the principle of least action that ideally helps to achieve this, because the principle of least action is the principle of the most economical action in terms of time and energy consumption. A company that operates on the principle of greatest action can only do so if it follows the principle of least action, after it has decided on its concrete goals positioned in space and time on the basis of the principle of greatest action. At the level of Nature, this means that the principle of least action is a necessary condition for physically realizing the biological aims specified in accordance with the principle of greatest action. This argument illuminates the *cosmic necessity* of the existence of the principle of least action. The necessity of the principle of least action involves, as a consequence, that the laws of Nature are eternal and invariable.

Unexpectedly, and without any effort to do so, the explanation of the hitherto seemingly insoluble question of the origin of physical laws fell into our hands. They arise as a necessity of the principle of the greatest action, as an integral part of it.

The Laws of Nature Exist in Reality and Have a Causal Role

There is currently no universally accepted criterion for existence. There is, however, one criterion that is self-evident. Samuel Alexander, the Australian-born British philosopher, has a famous dictum: "To be real is to have causal power" (quoted in Kim 1993, 348).[4] Natural laws are the sources of interactions, forces, changes—laws of change. Without natural laws, there would be no interactions, and there would be no changes. This means that without natural laws there would be no objects, since they are the manifestations of interactions, and no causes, since causes are the sources of change. In universal Reality, *natural laws are the source of causes, in the same way that interactions are more fundamental than particles* (Einstein and Infeld 1938, 313; Weinberg 1995). Natural laws are the cause, phenomena are the effect. Without the real existence of natural laws, the forces that hold molecules and atoms together would not exist, and thus no object would exist.

The source of physical change is force, the source of biological change is motivation, and the source of noological change is thought. The fundamental disciplines correspond to fundamental types of causes. The fundamental types of causes are physical, biological, and noological causes. There are no other fundamental types of causality. This corresponds to the fact that in natural science there are only three fundamental disciplines, related to the three fundamental principles of Nature.

Consider that life is first and foremost a ceaseless action, but also a myriad of possibilities for action that allows it to elevate itself over an almost unfathomably long period of time, by unfolding and fulfilling its highest potential. *Modern science has been unable to take into account the part of Reality without which life would not exist and would not be possible, namely the instinct for life. It is therefore of immense importance that it has become scientifically and philosophically understandable through the Bauer principle and its most general form, the principle of life.*

Nowadays, it is generally believed that life is only about survival. The discovery of the life principle tells something very different. The life principle tells us that life is to be directed toward a high quality of life. *Life is primarily directed toward feeling ourselves well, high above the level of mere survival, and this ensures the prospects to feel ourselves well, or better.* Our feelings are central to our quality of life. We conceive our instinct for life by our natural, instinctive feelings. Above all, life is about feeling good, individually and, more importantly, in our relationships with our families, our social communities, and Nature. The life principle is the principle of good feelings. The dignity of life is that every living being has a nature-given right to feel itself well, to feel good in the present, and good or better in the future. And that is to be respected fully, to the greatest extent. Without the life principle, the Universe would be completely inanimate. Without the life principle, rational mind cannot exist. Without the life principle, the Universe would be completely empty of meaning and values. It is the life principle that is the greatest treasure of the Universe, the source of all values. Its presence within us is the highest treasure of our life, and we are naturally inclined to pay attention, nourish, and care for this cosmic treasure in all its manifestations. We are most fundamentally and most personally interwoven with each other and with every living being through the life principle. This is why, if we appreciate our life instinct, we should appreciate the same life instinct in all its manifestations. *This recognition is the basis and essence of an environmental philosophy that values living communities* (Patterson 1998).

Life without its own natural law could not be fundamental and could not be an independent Reality. For without biological laws, life necessarily would become an appendage of matter; it would lose its dignity and intrinsic value, because without its own law, it would be subservient to physical laws and conditions. To recognize the Reality of the life principle is to recognize that life is an autonomous Reality having its own characteristics, following its own values, and with its own possibilities that are different from those of physics and incomparably greater, and all of this is due to the fact that it has its own independent principle.

We All Need to Know What Reality is, Since Our Future Fundamentally Depends on Our Understanding of Reality

There is still no consensus as to what philosophy is all about. To clarify what philosophy is, it is important to see that, at least as a matter of human nature, all human beings need a correct knowledge of Reality, and first and foremost a knowledge of the overarching principles of Reality. The three main branches of such a knowledge of Reality are the knowledge of life, the knowledge of the world, and the knowledge of humankind. And if philosophy is the study of the most general relations between these, then it is understood as a way of acquiring the most general knowledge of vital importance to all human beings. This philosophy, which is vital for every human being and seeks to show the most general interrelationships between life, the Universe, and humankind, can be called general philosophy. When this general philosophy reaches the knowledge of the most general principles, it becomes capable of informing humankind about the fundamental nature of life, the Universe, and humankind, thus providing a sound basis for a healthy culture.

For each of us, it is vitally important to understand general philosophy. It is also in our personal interest to make this general philosophy as simple, transparent, and effective as possible. Regarding the vital importance of our human relations for our full health, this philosophy has to be made understandable and easily applicable to everyone. We must therefore ask: What are the most effective tools of this general philosophy?

The Two Most Effective Tools of General Philosophy: Outlook of Life and Worldview

The two fundamental branches of general philosophy are the study of life and Nature. Humankind is living, fundamentally, in Nature. The nature-given life instinct sets the goals, and Nature provides the means to achieve them. *First of all, we need a clear, correct, adequate perception and understanding of the life instinct in a way that represents a suitable basis for our fundamental decision-making system. Second, we need to know and understand the fundamental constituents of Nature.*

If humankind acquires an essentially complete knowledge of the full explanatory systems of these two main branches of Reality, all humans will be able to shape their basic decision system accordingly. The explanatory system of life is called an outlook of life, the explanatory system of the world is called a worldview. *The two most effective tools of general philosophy are the outlook of life and the worldview.*

For humankind to live in harmony with the life instinct and with the Universe, it is necessary that general philosophy should be a unified system of knowledge. For this, the outlook of life and the worldview must be in harmony. This is only possible if the outlook of life is cosmocentric and the worldview is fundamentally life-based. Note that by the concept of the Universe we mean the unified whole of all that exists. The harmony of the outlook of life and the worldview is only possible if the Universe is alive. This means that we all need a life-centered worldview.

Civilizations Are Based on Their Worldviews, Which Determine Their Decision-Making

Worldview is defined as the way we conceive, that is, interpret and evaluate the world, fundamentally Nature. This is why "the intellectual level in every age has adapted to the generally accepted theories of the Universe" (Couderc 1964, 8, tr. AG).

An adequate worldview is the main goal and the most effective means of philosophy. It becomes even more effective when we recognize its basic type, which is determined by the ontological principles it takes into account and the relative weight attributed to them. The predominant worldview type determines the basic types of civilizations, their mindsets, their intellectual horizons, and their intellectual power. The best-known types of Western worldviews are the various appearances of materialist,[5] idealist,[6] and the dualist worldviews, the latter of which see both matter and consciousness as fundamental. In the light of the above arguments, all of these worldviews are fundamentally incomplete, because they ignore the most fundamental principle of the Universe, namely, the life principle.

All of our choices are ultimately determined by the way we interpret and value life and the world. If our mindset is, unconsciously, narrower than full Reality, we can still be consciously satisfied with the richness of choices available in our culture. Narrow mindsets alienate us from life and the world. Alienation is an ontological disease that destroys our relationship to the ontological basis of our personality and seriously diminishes our quality of life.

Knowledge is what offers humanity its greatest opportunities, and also its greatest dangers if it is inadequate or incomplete. There is no doubt that there is a physical aspect of life, and the physical sciences, including biophysics, have made significant strides in understanding these. But life, and it is Ervin Bauer who has done most to understand it, is not limited to its physical aspects. The central problems of biology (Szent-Györgyi 1970, 63), the nature of biological work and behavior, together with the central role of purposefulness in biology, cannot be understood by physics because its horizon is too narrow.

It is also important to see that materialism is a very profound metaphysical assumption about the nature of Reality. Since most people are unprepared to challenge this view to obtain a better foundation which is more profound and more powerful, the materialistic worldview so far tends to become, in practice, an unchallenged and unobserved magic-weapon of Western civilization, orienting most people toward a materialistic way of life.

Obstacles on the Road to Changing the Fundamental Type of Our Worldview into the Comprehensive, Ecological Worldview

The Difficulties of Changing Worldviews

The subordination of physical laws to the laws of biology, indicated also by the derivability of the physical principle from the biological principle as one of its limiting cases, changes the basis of our conception of Reality. After all, the physicalist worldview is the basis for the popular idea that, for example, human beings are basically made up of atoms. But if we conceive of ourselves as consisting of atoms governed by the physical laws, we become alienated from ourselves. To regain our lost selves, we must understand that life is more fundamental than matter. In fact, we have to rethink everything, expand our whole system of thinking, and build it on deeper foundations. In other words, our perception of Reality has to improve, namely by recognizing that our personality is based on our image of ourselves. This image is based on our image of humankind, which is based on our worldview. Therefore we have to build our worldview on a solid, correct foundation.

Nowadays we live in a situation in which *we have to change the foundation of our view of Reality*. In this regard, it is as if the ground beneath us, which we thought was an immovable continent, turns out to be a relatively small ice floe drifting toward a fatal situation. Now we have to leave this too-small sheet of ice and jump over to a previously unobserved continent in our neighborhood, which until now has been completely shrouded in fog. Only now is the fog dissipating and the continent becoming visible.

We have to admit that this is not an everyday situation, but one that requires an unusual effort to deal with. We need a preliminary mapping of the newly discovered continent and careful planning of the jump. Understanding Bauer's biology requires us to take an extraordinarily unusual step, the like of which we have never had to take in four hundred years. It is this unexpected challenge that has made it extremely difficult to understand Bauer's theoretical biology to date. We all have a seemingly rock-solid understanding of Reality, and it seems all the more solid because most people have the same conviction, and it is also underpinned by the dominant, physicalist worldview. Moreover, the dominant materialistic view is rooted in an ontological depth that is virtually inaccessible to most people. Replacing the hegemonic, centuries-old, deeply ingrained, and thought to be unquestionably solid physical worldview with the newly arising, ontologically deeper biological basis for understanding the world indeed presents us with a uniquely unexpected and profound turn in the history of modern science.

The scope and power of the human spirit is given by the breadth and explanatory depth of its worldview. A comprehensive worldview is much broader and much deeper than a materialistic worldview. It broadens humanity's intellectual horizons and increases its intellectual power to an extent never before suspected. All the paradigm shifts of the past centuries have occurred within the physical worldview. This time it is not a paradigm shift of this kind. Rather, we are confronted with the necessity of a qualitatively larger change, namely a shift of worldviews. It is this change of worldview, which qualitatively enhances the worldview that is the reference point for our understanding of Reality, that is the greatest significance of the Bauer principle. It is this change of outlook, which goes far beyond any preconceived notions, that gives Bauer's work its greatest significance.

Misunderstanding the Importance of Fundamental Principles

Although more and more physicists recognize it, the mainstream of physics still does not admit the fundamental importance of the principle of the least action. The prevailing view is that the principle of least action is mathematically equivalent to the fundamental equations of physics. But this view is based on the fact that the principle of least action is the source of physical laws. However, the principle of least action also has important advantages over physical laws, such as a panoramic overview of the whole process under consideration. While the basic differential equations of physics determine the course of physical processes as small changes from the previous state, step by step, the principle of least action determines the physical process between the initial and the final state. To use an analogy, the fundamental equations of physics work like "common soldiers," but the whole process, the strategy, is determined by their "general," the principle of least action. And what has been overlooked so far is that since the fundamental principle of physics determines all of physics, the principle of least action can be regarded as the definition of physics. Since the principle of least action is the principle of minimal action, *physics is the science of inanimate matter*.

This is of extraordinary importance. For until now it has not been clear what physics is, whether it is a complete science of Nature or not; and until it is unclear what physics is, it is difficult to recognize what biology is, and therefore, what life is.

Jaegwon Kim, perhaps the most distinguished scholar in physicalism, has put the essence of physicalism in this way: *Physicalism implies that physical theory can, in principle, provide a "complete and comprehensive theory of the world as a whole*" (Kim 1993, 96, italics AG), rather than simply a 'physical field' understood as a limited part of the world. Nevertheless, it is easy to realize that biological processes are not *governed* by the principle of least action. We have found scientific evidence to disprove physicalism. With Ervin Bauer's demonstration of the existence of biology's own laws and fundamental principle, including the demonstration that the Bauer principle is independent of physics, physicalism can be considered to have been scientifically disproved. Considering that physicalism is the dominant scientific worldview of our time, this in itself is more than a remarkable achievement.

Even more importantly, it is clear from comprehensive biology that the fundamental principles of the Universe have a worldview significance. The biological principle determines the unified system of conditions even for the physical processes that take place in the living organism, in accordance with biological goals. This is the biological control of physical laws. It is called biological organization. In other words, by continuously modifying the input conditions of physical laws in a way suitable for biological purposes, biology turns out to be the control science of physics.

Compatibility with Quantum Physics in the Absence of a Biological Law of Nature Does not Imply a New Biology

A consistently biological, scientific understanding of life is impossible without knowledge of the fundamental principle of life. The theoretical biology founded by Ervin Bauer owes its extraordinary scientific and philosophical importance to one single fact: the discovery and mathematical formulation of the fundamental principle of biology, from which Bauer derived all the basic equations of biology, including the equations of metabolism, reproduction, growth, excitability, explaining all the basic phenomena of life (Bauer principle [Bauer 1967]). With the discovery of the life principle, obtained by further development of the Bauer principle, the life instinct became suddenly and unexpectedly comprehensible on an exact scientific basis. Furthermore, building the entire architecture of theoretical biology on a single, overarching fundamental principle can be paralleled with the similar explanatory scheme of physics and logic. Surprisingly, the fundamental domains of the Universe, matter, life, and mind fit together like pieces of a mosaic puzzle, unveiling the full architecture of the Universe. Remarkably, the principle of matter, the principle of life, and the principle of reason are to be seen as scientific principles that have an extraordinary worldview significance, enlightening the blueprint of Reality by the all-encompassing fundamental principles of all existence. For the first time in the history of natural sciences, a complete picture has been obtained about Nature by a comprehensive science covering all the fundamental branches of natural science. This discovery draws attention to the extraordinary scientific and philosophical significance of the three fundamental principles unifying and mobilizing the entire Universe. The worldview significance of the overarching fundamental principles refutes the view that the basic worldview shift needed for sustainability is possible without the discovery of a biological principle that is independent of physics.

Bauer's seminal work was published in Russian in 1935 and in Hungarian in 1967; it has not yet been translated into other languages and is therefore relatively little known. The "fate" of this most extraordinary book is also most extraordinary, because its study and understanding have been hampered by a particular set of historical, political, scientific, and ideological obstacles. Bauer's major work has been likened by many to Einstein's main goal, that is, the discovery of the single equation of physics encompassing the "essence of Nature," from which all physical phenomena can be derived. Aleksei A. Ukhtomsky, the world-renowned researcher of neuroscience and founding director of the Institute of Life Sciences at the Leningrad State University, dedicated a reprint of one of his papers to Ervin Bauer: "To the Einstein of Biology" (Bauer 2003).

Nevertheless, the interpretations of Bauer's work so far are arguably wrong. Typically, even his greatest admirer and reviver of Bauer's work, Boris Petrovich Tokin, Director of the State Institute of Biology, Moscow–Leningrad, was wary of taking all the basic biological laws to be derivable from Bauer's principle—simply, as Tokin himself pointed out, because he knew of no such thing in physics (Tokin 1965, 116). But, as we have seen above, the principle of least action is such a thing.

It is no wonder, then, that even such exceptional scholars as the famous sinologist Joseph Needham were unaware of the existence of the Bauer principle and of the ontological significance of the fundamental principle of physics. In his view, scientific progress since the nineteenth century has transcended mechanical materialism, and the modern new scientific theories represented by Darwin, Einstein, and Planck "necessitated the adoption of a more organic philosophy" (Needham 1956, 505), and "[m]odern science and the philosophy of organism, with its integrative levels, have come back to this wisdom [of the Chinese worldview]" (583, insertions in brackets AG).

J. Baird Callicott, a distinguished research professor of the Department of Philosophy and Religion Studies and the Institute of Applied Sciences at the University of North Texas, begins his study "Metaphysical Implications of Ecology" with the idea that philosophy is substantively informed only by the universal and foundational sciences (Callicott 1989, 59). As he continues, "[e]cology and contemporary physics, interestingly, complement one another conceptually and converge toward the same metaphysical notions" (59). He adds that "[a] consolidated metaphysical consensus thus appears to be presently emerging from twentieth-century science" (59). A similar view is expressed by the Chinese philosopher Wang Kun, a professor at the School of Marxism of Zhejiang Normal University, who writes: "China's organic naturalistic philosophy is consistent with the foundation of modern new scientific theory and philosophy" (Wang 2014, 201, tr. AG).

Those who see modern physics as a step toward an organic worldview are right in that the strict determinism of classical physics ruled out the existence of life, while the indeterminacy of quantum physics made physics *compatible* with biology. Professor Wang Kun is right; organic philosophy is, indeed, *consistent* with modern physics. But the unification of modern physics with organic philosophy requires something much more important, a fundamental principle of biology that enlightens the entire universe of biology, namely in the sense that it exists *beyond* the domain of quantum physics. Such a fundamental addition to natural sciences can only be achieved on the basis of the universal principle of life. It would be a mistake to assume that relativity and quantum physics are based on an organic conception of the universe. Both are based on the principle of least action, the fundamental principle of inanimate matter. A prerequisite for both a science-based organic cosmology and ecology is the emergence of an exact science which is derived from a life principle providing scientific proof that the Universe is alive.

It is no mere coincidence that many consider the scientific solution of the problem of what gives intrinsic value to Nature as the "Holy Grail" of environmental philosophy (Callicott 2012, 767). For in a lifeless Universe what we call 'life' cannot be present. An organic cosmology cannot be qualified as scientifically grounded if it is merely compatible with, but does not transcend, physical laws. *This cannot be emphasized enough: The existence of a biological law, independent of physical laws, is a prerequisite for life.* Bauer puts it as follows:

> We can say that living matter has no laws of motion of its own, and that the laws of motion of living and inanimate matter are essentially identical. In that case, if we are consistent, we cannot speak of living matter in general. However, in this way biology becomes nothing more than applied physics and chemistry, that is, the application of physical and chemical laws to the complex systems we call living things. (Bauer 1967, 18, tr. AG)

Without laws, we cannot talk about theoretical science. It is the most general theoretical laws that make science so powerful. Only life with its own laws can give intrinsic value to the Universe. The task to put ecology on a scientific and cosmological basis requires us to take into account the leading role of the biological principle in the Universe.

The Cosmic Life Instinct Shows the Way to a Healthy Ecological Civilization

The prevailing view today is that it is *science,* implicitly identified with physics and confined to physics, that really matters, and that philosophy is of secondary or of no importance. It is important to see that science and philosophy are important in different ways. While the most effective instrument of science is given by the fundamental principles, the most effective instrument of philosophy is its worldview, and its most significant achievement is the creation and maintenance of civilizations through the development of a dominant worldview in society.

The great German philosopher Ludwig Busse (1862–1907) sees the entire history of philosophy as a series of attempts to establish a worldview (Busse 1917; cited in Péter 1939, 11). Accordingly, achieving a realistic, adequate worldview is the ultimate goal and fulfilment of philosophy. As David Naugle, who is a professor at Dallas Baptist University, writes in his book on worldview:

> A world "hypothesis," according to Freud, should be able to solve all problems, satisfy all interrogation, and put everything in its place. As an ultimate human ideal, a Weltanschauung [a worldview] in which one trusts should provide peace of mind or security by specifying the summum bonum and designating how to deal with life in practical ways. (Naugle 2002, 213)

In the social sphere, worldview is more important than science, because it fundamentally determines the perception of Reality, the way of thinking, the main direction of life paths, and the fundamental type of decision systems of the majority of people—and this can be understood most effectively in the light of comprehensive ontology.

Through institutionalized education, the fundamentally erroneous physicalist worldview becomes more and more deeply ingrained from generation to generation. The world corresponding to physicalism, which regards the inanimate phenomenal world as exclusive, exists only in our minds. Physicalist cosmologies describe the observable universe only in some respects and in broad strokes. Cosmological models of physics are unable to give account of the existence of even such a simple thing like a teapot, because they lack the complexity input of such details necessary to do so. But it is precisely in the richness of such complexities and fine details that life can unfold; and many little streams make great rivers, especially on a cosmic scale.

The physicalist worldview is "sick" because it is fundamentally incomplete. This is a fatal circumstance, because this fictitious worldview determines the main goal of modern societies, which consists of increasing material power and wealth regardless of all other considerations. It is this fictitious, one-sided worldview that determines most people's mindsets and basic attitudes toward their nature-given selves, each other, and their natural environment. It is this mind-narrowing worldview that determines the fundamental system of values in human societies. The almost incomprehensibly huge social changes that have taken place since the Middle Ages could not have happened without the social dominance of the new, physics-related worldview. In this context, it is clear that Bauer's theoretical biology lays the foundation for a fundamentally new era of a civilization based on a healthy, comprehensive, and life-principled worldview (Budapest Centre for Long-term Sustainability 2021).

The three fundamental principles, together with biological autonomy, can be referred to as the essence of the Universe. Learning and understanding their roles in the comprehensive worldview could serve as the basis for the general education necessary for the next generations to be able to build a healthy civilization. Since the basic type of worldview determines the basic type of socially dominant decision-making systems that dictate the basic type of civilizations, a comprehensive and healthy worldview allows for the foundation of a comprehensively healthy civilization.

Matter is only a tool in humankind's hands to achieve a happy, meaningful, and fulfilling life. This goal cannot be achieved without unceasing inner striving, without mobilizing our best abilities in a purposeful way, taking into account the comprehensive, communal nature of life. In order to improve our overall quality of life, we all have a personal stake in improving the quality of life in our communities and our environment. The more fully one lives in harmony with the cosmic life instinct, the closer one comes to the fullness of life, that is, to becoming a fully living being and to becoming "fully human." And since the life instinct embraces the whole of the Universe, living in harmony with the life instinct means a healthy relationship with Nature. Let us add that the ultimate meaning of life can only be ensured by conceiving the living and rational Nature of the Living Universe and by developing an attitude that establishes a good relationship with Nature. And this requires continuous emotional and intellectual development.

Taking into account the communal, social, and cosmic nature of our quality of life links our destiny to sustainability. Likewise, social sustainability requires a comprehensive worldview that prioritizes the value of life over the value of matter, while recognizing the value of matter and reason (Budapest Centre for Long-term Sustainability 2021). Sustainability requires a generation growing up in a culture that prioritizes the harmony between individual communal and cosmic values of life over an attitude of "profit above all else."

The greatest turns in science since Copernicus have all brought paradigm shifts within the physics-based, or rather, the physicalist worldview. The work of Ervin Bauer and the generalization of the Bauer principle to the natural motives of life represent the first attempts to go beyond this outdated worldview, namely *to establish a deeper, biological worldview, and thus to represent a major advance, even greater than that of the Copernican turn, in our understanding of the structure and nature of the Universe.*

The Life Instinct is an Essential Guide for All Aspects of Life

The cosmic life principle permeates all living beings. This life principle, when conceived from a personal perspective, can be called the *life instinct*. This instinct for life is what makes the life of living beings possible. The life instinct is so fundamental to all aspects of life that it has always played a fundamental role in all societies. Only in Western civilization has it been relegated to the background, but its myriad social roles have been preserved under different names and reinterpretations.

The content of the scientific formulation of the life principle makes it clear that the life instinct is not directed to mere survival but to *well-being and improvement of the comprehensive quality of life* in all its aspects.

Through the concept of the life principle given above, we can see the essence of life as the motivating 'urge,' 'intuition,' 'hint,' or 'inner impulse' that drives every living being at every moment to do everything possible to improve its life. This function of the life principle corresponds to the ancient concept of *'life force.'* The life force is manifested in the power that *heals* our wounds, propels the *development* of our organism since our conception, and *regenerates* our organism. The life force is the force of Nature which, acting within our organisms, inspires us to deep and true happiness, to *fulfil* and *flourish* our best physical, mental, and spiritual capacities, to create the most wonderful life we can.

In the literal sense, we live only when we live with our life force. That is, we live only to the extent that we live with our *full life force*. Every blade of grass has the capacity to live with full vitality from morning to night, day after day, for its entire life. If we do not want to lag behind them, we must do the same.

The life principle connects us to the *Living Universe*. In different cultures and different epochs, it may have different names like the cosmic life force, Mother Nature, *Dao*, *qi*, *rta*, *dharma*, Big Spirit, Supreme Being, or God in the sense of natural religion. Some cultures put emphasis on the life principle, such as the Chinese; others emphasize the aspect of decision-making, or biological autonomy, as Western civilization does. In full recognition of Reality, the primary role can be attributed to the life principle, since it is the directive power of the Living Universe. Knowing and keeping in mind the primary role of the life principle encourages us to nurture that connection. This role of the life principle is the same as that of those religions the essence of which is to cultivate the relationship between man and God. Its role as the supreme value, compass, and guide corresponds to the concept of the *Dao*, the Way. The life principle is the *path of life* as it is given by Nature.

The life principle connects us to "higher-than-individual," communal, and cosmic levels of life's organization. It corresponds to the role of *spirituality*.

The life instinct is the strongest *bond* that binds us to life. Our deepest feelings come from the life instinct. The feelings that connect our being most deeply to life can be called true *love of life*. The instinct for life is the source, the real basis of our deepest feelings for our family, for our country, for humanity as the family of all nations, for Nature, and, ultimately, for the Universe, the unified whole of everything that exists: Reality.

For a while after we are born, our response to everything that happens to us is in total resonance with the life instinct. Natural life is to live according to the values of the life instinct. It is the life instinct that sets the primary goals for our personal development, to improve the quality of life in all of its aspects. The life instinct is the real basis of our natural *self-identity*. Ultimately, our true, ultimately real self is the unity of the life principle and biological autonomy.

The life instinct is the same in all living beings. In our deepest and most essential self-identity, we are identical to each other. All of us are instinctively inspired to govern our life according to the cosmic life principle to make life the best it can be on this planet. This is the basis of natural *morality*, suggesting that we must act for the betterment for all living beings. Life is most beautiful in the comprehensive harmony of all life.

The cosmic scope of the life instinct is the source of respecting all life, of loving, appreciating, and respecting Nature. This is the basis of the *ecological* worldview.

Soul and reason are our treasures given by Nature, prompting us to achieve the aims given by the life instinct on Earth. The world provides the possibilities for realizing the life instinct. Since the life instinct is directed to the improvement, happiness, fulfilment, and beauty of universal life, the natural task of soul and reason is to discover the best possibilities of realizing these ends. This natural aspiration is the source of the desire to know life and the world, to discover life and the world as fully as possible. It is the source of innate curiosity, of the *desire to know*. The instinct for life is the source, the driving force, and the compass for understanding life and the world. It is the stimulus to develop philosophy and science so that both become complete, simple, and transparent.

Cosmic life brings itself to life; it is the source of *creativity*. Creativity in *art* is not without the guiding influence of the life instinct. There always have been, and always will be, composers, poets, and writers who wish to create the most beautiful pieces of art, and scientists and philosophers who strive to create the best theory.

Life's self-creating, self-fulfilling, and self-perfecting striving, this "positive feedback" of life, is the basis of the self-reflexive nature of *human consciousness*, of not only knowing, but of knowing that we know.

Ultimately, *the quality of our lives* depends on us, on our mentality, on our attitude, on our outlook. We can most fundamentally improve the quality of our lives if we have a correct, deep, and comprehensive worldview that guides our basic *decision-making* system.

It is the natural task of our soul, that is to say, of our unified and organized world of feelings, based on our life instinct, to conceive the life instinct as faithfully and as fully as possible. Our soul can conceive of the life instinct in an authentic way if our soul is clean, uncorrupted, and maintains its natural integrity. It is the natural task of the human intellect to comprehend the world as clearly and completely as possible. It is the natural task of the self, on the basis of our soul and intellect, to recognize the possibilities offered by the world that best correspond to the values of the life instinct, to make the *best decision*, and to make it come true.

This list is far from complete. It is natural that in all areas of life, even in modern society, the directive influence of the life principle is felt. But it is unnatural when the mainstream of society, media, science, philosophy, and culture turns away from the values of the life principle and hands over the leadership role to the principle of matter. Today, the conditions are ripe for a comprehensive, life-principled approach to emerge. The understanding of a healthy, life-principled, community- and nature-centered worldview is now within everyone's reach, thanks to the advancement of science to the level of comprehensive science.

The practical relevance of all these considerations is further underscored by the fact that in today's world, education and the "nervous system" of society, namely, the mass media, have increasingly taken the lead in child rearing. In recent decades, the Western school system and the Western media have played an increasingly important role in the life of every country. This change prepares the ground for the further strengthening of the dominant role of the materialist(ic) approach. So far, there seems to be a lack of awareness that this practice is increasingly ingraining a physicalist, materialistic, and nihilistic worldview into current and future generations. If this trend is to be reversed, namely for the sake of the development of a healthy civilization, it is essential that education and the "spirit" of the mass media are based on a healthy world view.

Notes

1. The term cosmic law designates the principle or set of principles believed to represent the most generalized Nature of the order of things in the Universe. Confidence in the existence

of a principle of order in the Universe at large is, in turn, reflected in a common belief that individual events within society are not random (hence meaningless) occurrences but parts of larger meaningful patterns that extend through time. Furthermore it is this confidence—that the entire Universe is established upon and governed by a principle of natural and moral order—that enables human beings, individually and collectively, to deal effectively with intellectual, moral, and spiritual life crises (Long 1995, 88).

2. This is also indicated by the fact that the entire Hindu religious system is based on the idea of *dharma*, literally the Great Order, the eternal law of the world. This law of Nature is the guideline of human law and justice, the spiritual principle behind the visible material world that floods the entire Universe, together with cosmic space, with life. *Dharma* is believed to be universal, forming the basis of all the laws of Nature. It is the principle of attraction and repulsion, the control of chemical processes, as well as the law of cause and effect, and logical thinking. From the metabolism of the simplest living beings to the beating of the human heart, *dharma* is the controller of life processes. Similarly, in ancient Chinese philosophy, *qi* 氣, the principle of life, is "the ultimate foundation for the existence of the universe" and the "ontological source of the universe" (Zeng 2011, 100). One of the oldest Chinese books, the *Yijing* 易經, the *Book of Changes*, says the Universe is based on the principle of life (Li 2008, 270).
3. A premise is a previous statement or proposition from which another is inferred or follows as a conclusion (*Oxford English Dictionary*).
4. Physicalism is equivalent to the claim that only physical causes exist. This claim rejects the existence of biological and mental causes (Kim 1993). This is contrary to our everyday experience that we can act according to our will. About the rebuttal of this claim see also Griffin (1998).
5. Today's dominant materialism is physicalism; it is defined fundamentally by the assumption claiming "the causal closure of the physical domain" (Kim 1993, 280), in accordance with "the claim that the world is fundamentally a physical world governed by physical law" (Kim 1993, xv). I use the term 'materialist' in the sense of 'matter-principled,' that is, the erroneous view that takes physics, the science of inanimate matter, to be the one and only fundamental science of Nature. The cornerstone of this matter-principled physicalist worldview is the assumption claiming that only one type of cause exists in the Universe, the physical ones.
6. Idealist worldviews assume that Reality is based most fundamentally on consciousness.

References

J.D. Barrow, F.J. Tipler, *The Anthropic Cosmological Principle* (Oxford University Press, Oxford, 1986)

E. Bauer, *Elméleti biológia* [*Theoretical Biology*; first published in Russian in 1935] (Akadémiai Kiadó, Budapest, 1967)

M.E. Bauer, M.Э. Бауэр, *Воспоминания обыкновенново человека* [*Memories of an Everyday Man*] (Асспин, Петергоф, 2003)

R.F. Baumeister, Self-control – the moral muscle. The Psychologist **25**(2), 112–115 (2012)

B.S. Bhante, *A sintoizmus* [*The Shinto*]. A világ nagy vallásai (Gondolat, Budapest, 1990)

Budapest Centre for Long-term Sustainability (BC4LS) (ed.), 95 theses for long-term sustainability: long-term sustainable, comprehensive and life-centred world order (Global Debates 4) (2021), https://bc4ls.com/95-theses-for-long-term-sustainability/. Accessed 1 Jan 2023

J.B. Callicott, The metaphysical implications of ecology, in *Nature in Asian Traditions of Thought: Essays in Environmental Philosophy*, ed. by J.B. Callicott, R.T. Ames, 51–64 (SUNY, New York, 1989)

J.B. Callicott, Intrinsic and instrumental value, in *Encyclopedia of Applied Ethics,* ed. by R. Chadwick (Elsevier, Amsterdam, 2012), pp.760–768

J.J. Clarke, *Nature in Question: An Anthology of Ideas and Arguments* (Earthscan, London, 1993)

J. Coopersmith, *The Lazy Universe: An Introduction to the Principle of Least Action* (Oxford University Press, New York, 2017)
P. Couderc, *A csillagászat története* [*The History of Astronomy*] (Gondolat Könyvkiadó, Budapest, 1964)
V. Csányi, *Az emberi természet* [*Human Nature*] (Vince Kiadó, Budapest, 1994/1999)
P. Davies, *Superforce: The Search for a Grand Unified Theory of Nature* (Touchstone, New York, 1984)
P. Davies, *The Fifth Miracle: The Search for the Origin of Life* (Penguin Books, London, 1998)
P. Davies, When time began. New Sci. (published Oct 9) **2468**, 4–7 (2004), https://www.newscientist.com/article/mg18424684-700-when-time-began/. Accessed 5 Feb 2023
A. Einstein, L. Infeld, *The Evolution of Physics* (The Scientific Book Club, London, 1938)
A. Grandpierre, Biological extension of the action principle: endpoint determination beyond the quantum level and the ultimate physical roots of consciousness. Neuroquantology **5**(4), 346–362 (2007). https://doi.org/10.14704/nq.2007.5.4.143
A. Grandpierre, Genuine biological autonomy: how can the spooky finger of mind play on the physical keyboard of the brain? in *ATINER's Conference Paper Series*, article PHI2012–0197 (2012). http://www.atiner.gr/papers/PHI2012-0197.pdf. Accessed 24 Nov 2022
A. Grandpierre, The origin of cellular life and biosemiotics. Biosemiotics **6**(3), 421–435 (2013). https://doi.org/10.1007/s12304-013-9173-9
A. Grandpierre, Biologically organized quantum vacuum and the cosmic origin of cellular life, in *Phenomenology of Space and Time: The Forces of the Cosmos and the Ontopoietic Genesis of Life: Book One*, ed. by A-T. Tymieniecka. Analecta Husserliana, vol. 116 (Springer, Cham, 2014), pp. 107–133.https://doi.org/10.1007/978-3-319-02015-0
A. Grandpierre, *Héliosz: a Nap és az élet új nézőpontból* [*Helios: The Sun and Life from a New Viewpoint*] (Titokfejtő Könyvkiadó, Budapest, 2015)
A. Grandpierre, The fundamental biological activity of the universe, in *Eco-Phenomenology: Life, Human Life, Post-Human Life in the Harmony of the Cosmos*, vol. 121, ed. by W.S. Smith, J.S. Smith, D. Verducci. Analecta Husserliana (Springer, Cham, 2018), pp. 115–140
A. Grandpierre, *The Ancient History of the Silk Road: The Role of Hungarians in Eurasia* (Titokfejtő Publishers, Budapest, 2021a)
A. Grandpierre, *Az Élet könyve. Az átfogó életfilozófia alapjai* [*The Book of Life: Foundations of a Comprehensive Philosophy of Life*] (Titokfejtő Könyvkiadó, Budapest, 2021b)
A. Grandpierre, D. Chopra, M. Kafatos, The universal principle of biology: determinism, quantum physics and spontaneity. Neuroquantology **12**(3), 364–373 (2014). https://doi.org/10.14704/nq.2014.12.3.747
A. Grandpierre, M. Kafatos, Biological autonomy. Philos. Study **2**(9), 631–649 (2012), https://digitalcommons.chapman.edu/cgi/viewcontent.cgi?article=1171&context=scs_articles. Accessed 19 Dec 2022
D.R. Griffin, *Unsnarling the World-Knot: Consciousness, Freedom, and the Mind-Body Problem* (University of California Press, Oakland, CA, 1998)
L.J. Henderson, *The Fitness of the Environment: An Inquiry Into the Biological Significance of the Properties of Matter* (The Macmillan Company, New York, 1913)
B. Hepburn, H. Andersen, Scientific method. The Stanford Encyclopedia of Philosophy Archive (2021), https://plato.stanford.edu/archives/sum2021/entries/scientific-method/. Accessed 22 Nov 2022
P.E.B. Jourdain, Maupertuis and the principle of least action. Monist **22**, 414–459 (1912). https://doi.org/10.5840/monist191222331
B.E. Kártyikné, *Lélegzünk és létezünk: Egyedi jóga a mindennapokban* [*We Breathe and We Are: Individual Yoga in Everyday Life*] (Magyar Természetgyógyászok Tudományos Egyesülete – Sanitas Egészségkultúra Központ – Sanitas Természetgyógyászati Alapítvány, Kiskunfélegyháza, 1989)
J. Kim, *Supervenience and Mind: Selected Philosophical Essays* (Cambridge University Press, Cambridge, 1993)

F. Lenormant, *Chaldean Magic: Its Origin and Development* (Samuel Weiser, York Beach, ME, 1999)
D. Lewis, *Philosophical Papers*, vol. 2 (Oxford University Press, Oxford, 1986)
C. Li, Fang Dongmei: philosophy of life, creativity, and inclusiveness, in *Contemporary Chinese Philosophy*, ed. by C.-Y. Cheng, N. Bunnin (Blackwell, Malden, MA, 2008), pp. 263–280
J. Lips, *A dolgok eredete* [*The Origin of Things*] (Gondolat, Budapest, 1962)
B.J. Long, Cosmic law, in *The Encyclopedia of Religions*, vol. 4, ed. by M. Eliade et al. (Macmillan, New York, 1987), pp. 88–94
J. Lovelock, *Gaia: A New Look at Life on Earth* (Oxford University Press, Oxford, 1987)
A.C. MacIntyre, *First Principles, Final Ends and Contemporary Philosophical Issues*. Aquinas Lecture (Marquette University Press, Milwaukee, WI, 1990)
D. Naugle, *Worldview: The History of a Concept* (W. B. Eerdmans, Grand Rapids, MI, 2002)
J. Needham, *Science and Civilisation in China*, vol. 2: *History of Scientific Thought* (Cambridge University Press, Cambridge, 1956)
F.S.C. Northrop, *Science and First Principles* (Macmillan, New York, 1931)
A.I. Oparin, *Life: Its Nature, Origin and Development*. Translated from the Russian by Ann Synge (Oliver and Boyd, Edinburgh, 1960)
J. Patterson, Respecting nature: a maori perspective. Worldviews Glob. Relig. Cult. Ecol. **2**(1), 69–78 (1998). https://doi.org/10.1163/156853598X00064
Z. Péter, *Világnézettani alapvetés* [*Foundations for a Science of Worldview*] (Szerző kiadása, Debrecen, 1939)
C. Riviere, Soul, in *Encyclopedia of Religions*, vol. 13, ed. by M. Eliade et al. (Macmillan, New York, 1987), pp. 426–427
R. Rosen, *Optimality Principles in Biology* (Springer, New York, 1967)
G.G. Simpson, Biology and the nature of science. Science **139**(3550), 81–88 (1963). https://doi.org/10.1126/science.139.3550.81
M. Stöltzner, The principle of least action as the logical empiricist's *Shibboleth*. Stud. Hist. Philos. Mod. Phys. **34**(2), 285–318 (2003). https://doi.org/10.1016/S1355-2198(03)00002-9
D. Sturm, Natural law, in *The Encyclopedia of Religion*, vol. 10, ed. by M. Eliade et al. (Macmillan, New York, 1987), pp. 318–324
A. Szent-Györgyi, *Egy biológus gondolatai* [*Thoughts of a Biologist*] (Gondolat, Budapest, 1970)
B.P. Tokin, *Az elméleti biológia és Bauer Ervin munkássága* [*Theoretical Biology and the Oeuvre of Ervin Bauer*] (Akadémiai Kiadó, Budapest, 1965)
P. Tompa, G.D. Rose, The Levinthal paradox of the interactome. Protein Sci. **20**(12), 2074–2079 (2011). https://doi.org/10.1002/pro.747
K. Wang, 王锟,《怀特海与中国哲学的第一次握手》[*Whitehead's First Handshake with Chinese Philosophy*] (北京大学出版社, 北京, 2014)
S. Weinberg, Reductionism redux. The New York Review, published 5 Oct (1995), https://www.nybooks.com/articles/1995/10/05/reductionism-redux/. Accessed 18 Jan 2023.
W. Yourgrau, S. Mandelstam, *Variational Principles in Dynamics and Quantum Theory* (Sir Isaac Pitman and Sons, London, 1955)
A. Zee, *Fearful Symmetry: The Search for Beauty in Modern Physics* (MacMillan, New York, 1986)
Z. Zeng, Semantic criticism: the 'westernization' of the concepts in ancient chinese philosophy: a discussion of Yan Fu's theory of Qi. J. Artic. Front. Philos. China **6**(1), 100–113 (2011). https://doi.org/10.1007/s11466-011-0127-5

Attila Grandpierre (Hungary) is an internationally renowned solar astronomer, quantum physicist, biologist, cultural historian, musician, and poet. He was a senior research fellow at the Konkoly Observatory for many decades. At present he is the director of the Hungarian Ecological Civilization Research Institute and the research president of the Budapest Centre for Long-Term Sustainability.

Holographic Philosophy as a Philosophical Basis of Cosmic Life Theory

Chunyou Yan

Abstract Holographic philosophy was first introduced in China during the 1980s. It was initially referred to as 'theory of cosmic holography' and later renamed 'theory of cosmic holographic unity.' Its main proposition is that parts contain all the information of the whole. In the first section of this chapter, the general process of the emergence of holographic philosophy in Chinese discourse is described. The second main section provides an analysis of the basic principles of holographic philosophy and points out its provability. The universe is a vast living entity, and every part of it also contains life, which is the inevitable conclusion of holographic philosophy. The third section provides further explanations regarding the principle of holographic philosophy. From this point of view, all things are to be considered as aspects of cosmic life at different stages of evolution, and they are all evolving towards higher states of life. In the fourth and final section, the high significance of the holographic theory is discussed in more detail and in terms of the whole–part relationship.

Keywords Holography · Holographic philosophy · Parts and whole

Overview of Chinese Holographic Philosophy

The first Chinese paper on holographic philosophy was "The holographic theory of the universe" ("Yuzhou quanxi lun 宇宙全息论") by Wang and Yan (1985). In this paper, we proposed the concept of a 'cosmic holographic principle,' which was defined as 'every part of the universe contains all the information of the universe as a whole.' In "The holographic theory of mind" ("Jingshen quanxi lun 精神全息论") (Wang and Yan 1986), we applied the holographic principle to analyze the structure of the human mind. The article "The cosmic law of holographic reiteration" ("Yuzhou

Chunyou Yan—Translated by Bin Yang and David Bartosch.

C. Yan (✉)
College of Education for the Future, Beijing Normal University, Zhuhai 519087, China
e-mail: 86167@bnu.edu.cn

School of Philosophy, Beijing Normal University, Beijing, China

D. Bartosch et al. (eds.), *Towards a Philosophy of Cosmic Life*,
https://doi.org/10.1007/978-981-99-2131-7_5

quanxi chongyan lü 宇宙全息重演律") (Wang and Yan 1987a) proposed that the development process of any subsystem of the universe is a holographic reiteration of the process of system generation. The "The mental law of holographic reiteration" ("Jingshen quanxi chongyan lü 精神全息重演律") discussed the phenomenon of holographic reiteration with regard to the human mind (Wang and Yan 1987b). In the paper "The theory of cosmic holographic unity" ("Yuzhou quanxi tongyi lun 宇宙全息统一论") (Wang and Yan 1987c), the theory of cosmic holography was renamed ' theory of cosmic holographic unity.'

In 1988, two books were published: *The Theory of Cosmic Holographic Unity* (*Yuzhou Quanxi Tongyi Lun*《宇宙全息统一论》) (Wang and Yan 1988a) and *The Science of Cosmic Unity* (*Yuzhou Tongyi Kexue*《宇宙统一科学》) (Wang and Yan 1988b). Here, the authors deeply and systematically explored the main issues and contents of holographic philosophy. *The Theory of Cosmic Holographic Unity* was reprinted in the years 1990, 1992, 1995, and 1998. In 1991, *The Theory of Cultural Holography* (*Wenhua Quanxi Lun*《文化全息论》) was published (Yan and Yan 1991). The work applied holographic philosophy to the study of cultural phenomena. *The Mystery of Mind: Holographic Noology* (*Jingshen Zhi Mi: Quanxi Jingshenxue*《精神之谜——全息精神学》) analyzed the human mental structure from the perspective of holographic philosophy (Yan 1991).

In 1996, "Outline of holographic philosophy" ("Quanxi zhexue lungang 全息哲学论纲") was published (Yan 1996). In this work, the relatedness of different philosophical systems to the holographic principle was discussed. The paper "The advantage of holographic theory over systems theory and reductionism" ("Quanxi lun dui xitong lun yu huanyuan lun de chaoyue 全息论对系统论与还原论的超越") addressed the relationship between holographic theory and systems theory and reductionism (Yan 2008a). The author also published "Introduction to holographic philosophy" ("Quanxi zhexue yinlun 全息哲学引论") (Yan 2008b) to analyze the development process of holographic philosophy.

Proof of the Basic Proposition of Holographic Philosophy

The only fundamental principle of holographic philosophy is that parts contain all the information of the whole. In fact, this opinion is not new at all. It has existed since ancient times, both in China and the West.

For example, traditional Chinese medicine (TCM), which has evolved since two thousand years ago, is related to this view in two ways. First of all, in TCM, each part of the human body is regarded as an epitome of the whole human body. For example, the acupuncture points on one's ear correspond to the whole human body, and each acupuncture point on the ear corresponds to an organ of the body. Thus the ear resembles a fetus turned upside down. Secondly, human beings and all things under Heaven and Earth are one. This is called the 'unity of Heaven and man' (*tian ren he yi* 天人合一): the sky and humans correspond, and the human being is a microcosm of Heaven and Earth.

There is a similar view in Buddhism, especially in the *Avatamsaka Sutra* (*Huayan-jing* 华严经). Mahayana Buddhism describes the whole universe as an interconnected cosmic network. There is such a metaphor of Indra's net in *Avatamsaka Sutra,* which is imagined as a huge net with gems hanging in the palace of Indra. You can see all the other gemstones reflected in each of its gemstones. Every object in the world is just like this: They are what they are by containing all other objects, and by actually being all other objects. "Countless Buddhas appear in every grain of dust." "Indra's net" permeating the whole universe symbolizes the infinite connection of everything with everything in the universe. Its horizontal line traverses through space, and its vertical line runs through time, creating each entity at their intersection. These entities are grains, that is, infinite "transparent" objects. The light of the "divine" shines and penetrates every grain, and the ghost of information reflects and reaches every object. Each grain reflects not only the light from each grain but also the light reflected by the whole universe. In essence, each object reflects not only the information from each object, but also the information reflected by the whole universe.

The theory of reality or *paramattha* in Tiantai Buddhism emphasizes 'three thousand realms in a single moment of life' (*yi nian san qian* 一念三千), which means that even a very short period of thought activity contains the whole universe. In a single moment of all living beings, there are all phenomena and all realms. And all phenomena and realms are inherent in a single moment. Every single thing in the universe is just like this. There is no need for them to rely on other things. Therefore, every single thing and all living beings are per se harmonious and self-sufficient, or in other words, they are omnipotent.

In the late 6th and early 5th century BCE, Alcmaeon, a Pythagorean philosopher and medical writer in ancient Greece, believed that man is a small universe and a microcosm. The human body is a reflection of the structure of the world. Giordano Bruno (1548–1600), an Italian Renaissance philosopher, believed that each fraction of matter contains other fractions, so that it contains the whole world, and each fraction of matter contains the world soul. The German thinker Gottfried Wilhelm Leibniz (1646–1716) developed the famous understanding of monadology. It holds that each monad reflects the whole universe—not only the current universe but also the past and future of the universe—and that every monad has feelings.

Therefore, the present author's view also means a sort of summarizing revision of similar views from ancient to modern times. What are the differences between this view and the previous ones?

First and foremost, we understand this kind of mutual inclusiveness from the perspective of information, that is, what is contained in each part of matter or the world is not the substances of things or the real things, but the information of these things. In other words, they are virtual rather than actual existences. Various previous viewpoints have often led to mysticism. The reason is that they did not follow the information perspective. Secondly, this new view was systematically and fundamentally expounded from the combined perspective of information and holography for the first time. Thirdly, in our view, the most fundamental aspect of holography lies in its intrinsic characteristics. In other words, every part of the universe or matter contains all the information of the universe or matter, not just, as it was widely promoted in the past, the mutual reflection and interaction between things. In this way, if every part of the universe continues to develop, it can evolve into a small universe.

How should we understand and prove this proposition? First of all, we should understand the concept of information. There are various different understandings of 'information,' and its meaning varies greatly in different disciplines. By information, we mean the sum of all attributes of being itself. These attributes involve the modality, form, structure, nature, ability, and all properties of the object. It can be seen that this definition is made from the perspective of ontology, and it refers to the ontological information of things.

At the same time, it is also defined from the perspective of epistemology. As covering 'all' attributes, it must contain the attributes that we do not know. Therefore, 'information' means the sum of known and unknown information. And the proposition that parts contain all the information of the whole means that parts contain all the known and unknown information of the whole. And this is an understanding from the perspective of human cognition. However, regarding the things in themselves, there is no distinction between known and unknown information.

This proposition can be further extended, that is, everything in the universe contains all the information of the universe. The universe is the largest entirety that we can describe, and everything else is part of it. This proposition features provability and non-provability.

Let's start with its non-provability. All propositions have their scope of application and their limits of provability, that is, they all feature non-provability, and so does this proposition. As this proposition involves the whole universe, it can be seen as an ultimate proposition. Since it is ultimate, it is difficult to prove. However, it can be seen as a self-explanatory proposition. And it is a last resort to "prove" a truth with 'self-evidence.' Human cognition always requires an endless chain of proofs. However, it is impossible to go on proving indefinitely, so we have to resort to 'self-evidence' to end the chain of proofs. The ultimate proposition of all theories is just like this—showing their non-provability. Its non-provability needs to be proved. And only upon that, a proposition can be completely proved.

In what sense is this proposition unprovable? Are there not a lot of scientific examples to prove it? There are indeed many people who have been able to unveil many practical and scientific examples of holography, such as holographic photography or biological holography. However, these are finite proofs. They can only prove the application of holography in some fields rather than in view of an infinite domain. In other words, they cannot prove holography in the sense of philosophy. These proofs at the scientific level cannot replace philosophical proofs.

The difficulty of proving this proposition lies in 'the whole.' If this proposition is changed to 'parts contain parts or limited information of the whole,' it can be proved by showing that parts contain a kind of information of the whole. However, 'the whole' cannot be proved unless 'the whole' is finite. If all the information of the universe is finite, this proposition will be confirmed scientifically sooner or later. However, a finite universe is inconceivable. How can we even imagine that the universe has a beginning and an end? And what is it before the beginning and after the end? Is it the universe as well? Similarly, it is not feasible to imagine something "beyond" the universe, because it would still represent a part of the universe. Therefore, the universe cannot be finite. However, the infinite universe is also inconceivable. We cannot imagine that there is no beginning, no boundary, and no end of the universe, and it is certainly quite frightening to think about this. In fact, the universe in our mind is always self-centered. In other words, the universe in our mind is still a finite universe, as indicated by the two Chinese characters *yu* 宇 (space) and *zhou* 宙 (time).

Since the universe can only be infinite, this proposition will indicate that the parts contain infinite information. The infinite universe must contain infinite information. It makes no sense to say that the infinite universe contains finite information. Otherwise, the proposition that parts contain all the information of the whole will never be proved empirically. If the parts do contain all the information of the universe, we can never identify all the information contained in any finite part.

If we make this important proposition the ultimate basis of a theory, but we cannot prove this proposition—isn't it absurd then? In that case, what is the point in making it the basis of further deduction? Unfortunately, not only ours but all theories are absurd. The ultimate basis of any theory cannot be proved. In this regard, the final principle, as Oswald Spengler (1880–1936) said, has the meaning of faith (Spengler 1988). This "belief" is certainly not unfounded. Unfounded belief is not faith, but wishful thinking. Most of it is based and coming from experience and unspeakable or unexplainable things beyond experience, such as aesthetic feelings, awe, mystery, and other experiences and feelings, which are the foundation of these ultimate principles or "beliefs," that is, the ultimate source of those principles. Aristotle's prime mover, the Christian god, Hegel's absolute idealism, Spinoza's concept of substance, Leibniz's pre-established harmony, Kant's free will, Schopenhauer's philosophy of will, Heidegger's "existentialist" philosophy, etc., are all such ultimate propositions with faith-related significance.

On the other hand, there lies a certain provability in this proposition. From a relative point of view, it can be proved:

(1) It can be proved from an indirect or negative point of view. It is important to note that if the universe would not be holographic, it would be incomprehensible; and as we can understand the universe in a certain sense, this indicates that it adheres to the holographic principle. If the universe were not to follow this principle, it would mean that the universe 'here' would be absolutely different from the universe 'there.' In this way, our understanding of the universe 'here' could not be applied to 'that' universe, and the universe 'there' would be unknown. If that were the case, any cognition would be impossible, because any cognition is not universal, and it represents only the current particular experience. Therefore, no human knowledge could be formed. This is not the case, and the opposite is true. This proves that the universe 'there' is unified with the universe 'here,' that is, it contains the same information.

(2) Also in perspective of the universe itself, one can provide a negative proof that the universe itself must also be holographic. If this were not the case, there would

be no order and regularity. If the universe were not holographic, the universe would be different everywhere. If it were different everywhere, there would be no unity, let alone the regularity and order which require the essential condition of unity. The existing universe represents strict regularity and order, thereby fully indicating the holographic character of the universe. In other words, there are holographic constraints and holographic inclusion of information among different parts of the universe and between the parts and the universe as a whole.

(3) It can be proved from the perspective of human cognition. Human cognition also follows the holographic principle. Specifically, it always infers the whole from the part and derives the unknown from the known, which indicates that the whole is implicit in the part and the unknown is implied in the known. Otherwise, such inference and cognition would not be possible. Newton did not measure the whole universe to discover the law of universal gravitation. In fact, after obtaining the data from a very small area here on Earth (which is just a "point" in the universe), he applied this data to the whole universe. All human cognition is like this. It proceeds from existing knowledge toward yet unknown knowledge and extends finite knowledge toward the infinite. This holographic principle of cognition is the ultimate premise of all cognitive activity. Otherwise, there could be no cognition.

(4) In this context, we can obtain a positive proof, and this proof is actually the self-evidence of the holographic principle. If there is holography in the universe, as long as there is a holographic phenomenon in the universe, we can prove the holographic nature of the universe as everything in the universe contains all the information of the universe. Therefore, both physical holography and biological holography can prove the existence of cosmic holography as there cannot be absolutely isolated phenomena in the universe. The existence of these holographic phenomena must have a universal basis. The nature of any special thing is the manifestation of the universality of the universe.

Therefore, experience itself is not a special something but has universality. Experience, especially learning through personal experience, represents a most particular aspect. However, there are still universal attributes in the most special facts. And that is the reason why universal knowledge can be produced from experience.

'Experience' does not mean ordinary experience here. Ordinary experience represents the activities of sense organs, such as seeing, listening, tasting, and touching. These activities certainly have their universality. However, only the people with a sensitive, clear, and wisdom-centered mind, rather than a dull one, can unlock a vast world from these superficial sensations. This is related to another topic of holographic theory, that is, the internality of holography. People often wonder why they cannot find that everything contains the information of the whole universe. The reason is that they look at the universe with their naked eyes rather than with their minds. The naked eyes are dull without sagacity.

The primary meaning of holography is not that of a universal connection, as some people say. If it were, there would be nothing new. The internality of holography means that the interior of each thing itself and the interior of each part of the universe contain the information of the whole universe. More specifically, they contain the information of all the properties of matter. This means that the internal information of anything and any part of the universe is equal to the universe. If it is allowed to evolve indefinitely, everything in the universe can be evolved from anything or any part of it, leading to the generation of a new universe. In this sense, each thing or part is a small universe, a universe of information.

Therefore, any part of the universe is not dead, but alive, and every part contains all the information of life. The universe itself is an eternal life, and everything is an integral part of this eternal life. What we call life is only a stage, a form, and a moment of this enormous life. All things are just in different life forms at different stages of evolution, and they are all evolving towards higher states of life.

The word 'life' is not a metaphor here. The universe is actually life (itself). The universe itself exists and moves forever. Its energy is so immense and unimaginable and its vitality is powerful. As we cannot imagine any so-called life that can compare with it, how can it not be life? In the process of eternal movement, the universe constantly changes its existing forms and is turning into one new thing after another. Which aspect of so-called life has such strong vitality? And which one has such a huge ability to change? As long as there is such an ability to change and transform everything among all things, it can feature eternity. The universe is such an eternal life. It is constantly regenerated from the death of all things. The death of things is just a bridge for the universe to transit from one life to another. The universe is forever regenerated from this eternal death.

It goes without saying that the information of the small universe or other things contained in the parts can emerge only under certain conditions. A rock can never directly become a pig. However, under certain conditions, it could. However, for a thing or a part to evolve into a real universe, it would require infinite conditions, and it would require infinite information to appear in infinite time and space. This is impossible, because the universe already exhibits this infinite condition. Therefore, individual things or parts can only have finite conditions.

For this reason, finite things or parts can only reveal a limited amount of information. Every thing tries to show more intrinsic information, and this ability is limited by its conditions. Therefore, they can only be revealed as specific things. When different kinds of things are arranged, each becomes a link in the whole cosmic chain, a link moving forward and trying to become a higher-level thing, each trying to become a whole and a universe. This forms a series of countless levels from low to high. The lower levels are limited by the higher, but cannot become a higher level. All levels are restricted by the universe as a whole. Viewed as a process, it can be represented in the sense of a theory of cosmic evolution.

Therefore, every being has infinity and contains infinite information. However, this infinity is potential in the small universe, that is, each being is only a potential infinity rather than a real infinity. It cannot become a real universe, but it inherits the possibility of infinite development, or this infinity is a potential in the finite thing. Therefore, every being is only a potential universe, that is, a potential infinity. The infinite information it contains is to be thought of in a similar way as potential information that is not being revealed.

Since each thing contains all the information of the universe, why can't any of it present all the information? This is not difficult to explain. Finite information can only be represented in finite time and space, and the information at a certain level can only be represented at that level. In addition, as there is the need to form a whole, the inhibition effect of the whole limits the display of all the information contained in the parts. This is just like our body. Although each organ and even every cell contains all the information of the whole body, as a part of the body it has to exist within the framework of a particular form and cannot develop indefinitely. Otherwise, instead of forming the whole body, it were to start a process of continuous disintegration. The same is true of human society. Everyone has the right to be president, but not everyone can reach that position. Otherwise, society would disintegrate and everyone would be a "country." Therefore, each part can only show some specific information, rather than all the information contained. If each part were unwilling to become a part of the whole, all the parts would eventually perish together. If a part were not part of a whole, it would also perish. And it could not be called 'a part,' because a real part necessarily belongs to a whole.

This internal or intrinsic aspect of holography cannot be fully verified by means of experience. The information contained in everything is infinite, which cannot be proven empirically. But we can prove it by theoretical deduction:

First, it is generally accepted that the transformation among things is both experienceable and universal. This transformation is possible, because the transformed things were in some way contained or implied in original things. Otherwise, transformation would not occur at all. Some people do not realize this and say that the emergence of a new thing means an essential change and that the new thing is not implied in a preceding "original" thing. If a new thing were not originally included in other things, this would mean that it were not derived from an original thing. Then where would it come from? The only possibility would be that it came from anywhere and anything and that anything could become anything. However, in fact, particular things can only transform into specific new things. This shows that this change of things involves its own intrinsic "logic," and a new thing must have been logically hidden in a (preceding) original thing. Otherwise, such transformation would not take place. Probably everyone will admit that this transformation process will not end and that it is infinite. Everything has the potential to achieve infinite transformation, indicating that it contains infinite convertible information.

Secondly, the universe features wholeness, that is, everything in the universe constitutes a co-integrated whole in which all parts are mutually regulated. All things in the universe are mutually regulated. Without this mutual regulation, the universe would not show such strict regularity. This mutual regulation makes every thing and every part integrally related to the whole universe, and every part is in omnidirectional connection with the whole universe. Through this mutual regulation, the information of all things in the universe is exchanged between different things. In other words, each thing receives the information of other things. The universe, as a whole, regulates everything. This regulation channels the information of the whole universe into every thing and every part. Without the input of information, it is impossible to regulate and restrict something. However, the universe is infinite. How does the infinite universe regulate a finite thing? We can only answer in this way: The infinite universe regulates infinite things with infinite power and infinite methods in an infinite manner. This regulation is far beyond our cognition, because no matter how small a part of everything is, such as a grain of sand and a particle, it is subject to this infinite regulation.

Needless to say, this is only our understanding and conjecture. There will be no final answer as to how this holographic nature actually arises. Holographic nature is infinite. There can be no definite answer to something that is infinite.

From the above, it is concluded that since every thing or part contains infinite information, it contains all the information rather than finite information. If it were to contain only finite information, it would not be 'holographic.' Many people argue that holography is incomplete, because we cannot get all the information about the whole from the parts. Since what we can see is only finite information, the essence of holography is 'similarity.' If this were true, the concept of holography would be meaningless. Holography must be total, and the information it contains has to be infinite, that is, (universal) holography is absolute.

Why is it absolute? Why and how do finite things contain infinite information or infinite cosmic information? This, by nature, is beyond our comprehension. But what is 'incomplete' here is not Being as such, but our cognition. 'Holographic incompleteness' is caused by our cognitive ability, rather than by Being itself. The so-called incompleteness is due to our limited cognition.

In terms of the things themselves, the "incompleteness" is caused by the inhibition that the whole universe and all other things exert on the thing or the part. The inhibition or restriction leads to the display of certain finite information or some specific information, while other information exists only potentially. As the infinite information is infinitely potential, we have no chance to receive it as such. What we see is only on a superficial or more shallow level of visible information. In fact, every thing or part has the ability to become a universe. If it were given infinite time and space, it could demonstrate infinite information. Every thing is potentially infinite and implies a vast world.

The information contained in the part is not dead but alive. It is just dormant, not dead. As soon as the day dawns bright, it will be awakened. The hidden information is ready to embark on an infinite journey at any time.

(5) Our position can be proven in view of the source of information. The basic principle established by holographic philosophy is that information cannot come from nothing, but only from Being. And this principle itself validates holographic philosophy in a very fundamental sense.

We can think about the following questions: Where do things come from in the universe? Where do new things come from? Are they absolutely new? In other words, do they suddenly appear without having existed in any form before? The essence of these questions is to know whether there is absolute creation in the universe, or whether new things already existed in other forms, and the so-called "new things" are just a manifestation of that. I think only the latter is more intelligible, while the former is inconceivable.

Where does the new information in a system come from? Some people believe that it comes from outside the system. In other words, the emergence of new things lies in the addition of new information from an outside of the system. We may admit this for the sake of further reasoning. However, according to this logic, we will arrive at the absurd conclusion that things in the universe can only come from outside of the universe. The universe is the largest system we can think of. If the emergence of new things were to depend on information from outside of it, we could only say that the emergence of things in the universe would result from an addition of new information from outside the universe. However, this is absurd, because the (infinite) universe has no "outside space."

This can be illustrated by taking a small system—Earth—as an example. Where do the organisms on Earth come from? According to the above-mentioned view, they result from the interplay of various factors outside Earth, without which no organisms would be possible on Earth. If our planet were a closed system, no life could have arisen on Earth. We can admit this. However, the fact that, up to now, no life has been definitely found outside the Earth so far indicates that inanimate factors outside Earth already contain the information of life, and they are merely activated on Earth. At the very least, an inanimate Earth itself and such factors operating on it already contain the information of life. If there were no implicit life information in these factors, no matter how many factors might be added from outside the system or how many interactions might occur—it would not be able to produce life.

Those lower levels of matter already contain various higher-level material forms or attributes. Otherwise, it would be impossible that higher-level material forms could evolve from these lower levels of matter. The evolving process from an inanimate planet to life on Earth shows that life has been intrinsic in those inanimate forms of matter. If no life information were contained in it, no matter how appropriate the conditions would be, it would be absolutely impossible for life to unfold.

By a logical extension of this point, any information of the universe can only come from itself. Things already existed in the universe in the form of information before they came into being. The generation of so-called new things is just a manifestation, not a creation at all. There is no absolute creation in the universe, and everything has always existed in some way. Since the existing things have existed in the past in some way, all things in the future must exist in the present in some way. Since our Earth—which began as a "giant rock"—was able to evolve life, this indicates that the information of life was already contained in the "stone" (as well). Otherwise, it would have been impossible for life to evolve on Earth. Since the rocks on Earth evolved into life, stones anywhere in the universe must also be able to evolve into life under the right conditions.

On the contrary, if we assume that there is absolute creation in the universe and new things have never existed in any form, the emergence of these things could only be counted as a miracle. We would have to ask: Where would they have come from? In that case, there would be no source for the information that they exhibit, which is inconceivable.

Therefore, the so-called development of things is essentially a process of unfolding, while the so-called disappearance is essentially a process of convergence.

Each thing now contains not only the whole history of its own evolution, but also the whole process of the future, and even the history and course of the whole universe. There is no real creation or disappearance in the universe. The flow of the whole universe is a process of continuous circulation of various modalities and forms.

Since information can only come from (this process of) Being, and since it all already has existed and exists in some way, all future things are contained in any aspect of matter, just as all things today have been contained in any matter in the past.

However, according to this human logic, we will arrive at a proposition like this: Everything in the universe is inevitable and there is no contingency. In other words, all things and their existing states and trajectories are predetermined. This is a problem proposed by Zhaoqiang Wang[1] at that time, and it includes the most powerful and in-depth question that I have ever encountered. Since the development process of all things in the universe is only the display of potential information, there is no contingency in the universe, and everything must be predetermined. Based on this, Zhaoqiang Wang proposed his question: Are there a minuscule Cunzhen Wang and Chunyou Yan in the original singularity of the universe? (Wang 1989) This might sound strange, but, in the sense of the logic of the holographic principle, it must be the case.

In fact, this question appears not only in holographic philosophy, but also in our general thinking. We admit that the universe is regulating (itself), and this regularity must be of inescapable necessity. What is necessary in this way must be necessarily so, and no change is possible. Similarly, we can raise a further question. After the destruction of the finite universe, will the next process be exactly the same as the previous one? If it is exactly the same, will there be exactly the same people and things in the second process as in the first process? For example, is it possible that there are still the people called Zhaoqiang Wang, Cunzhen Wang, and Chunyou Yan? And will this Zhaoqiang Wang still ask Cunzhen Wang and Chunyou Yan the same question? If so, it would seem absurd, but if not, then how can one say that the second process follows the same regularity as the first process? Then what order would there be to talk about?

Of course, one can come up with approaches to solve the problem. For example, one could say that the regularity is only qualitative, not quantitative; or that it is only categorical, not individual. The question posed by Zhaoqiang Wang could be answered in the same way. In holographic relations, all the information contained in parts refers to categorical or essential information, not individual or quantifiable information.

In this way, it is true that the above question will have an answer. However, we will have a new problem. If the regularity is like this, it is vague, indistinct, and general; and the things generated by it should also be vague. However, this is not the case. Every thing is more precise than our human mind can imagine. Science and technology reveal that it cannot be more imprecise than a hair's breadth, as in the case of the space shuttle for example. There are many other examples. If there would be some slight impreciseness in the expression of human genes, the individuals would be deformed. Our body, our life span, every hair, every cell, its mode of operation, and its relationship with other cells are predetermined, and there can be no mistakes. Each of our organs and their functions are also "prearranged," and there is no way there can be any confusion. It is beyond our comprehension how everything is predetermined, and how nature can display such superior wisdom.

The above answer also has its own intrinsic problems, because it runs counter to the meaning of the concept of holography itself. It implies that every being or thing would not contain all the information but only part of the information. A thing cannot be generated with only categorical and essential, or universal regulatory information. It can only be formed with *all* the (infinitely complex) pieces of specific information to make it 'this one.'

In this regard, each individual thing actually contains all universal information, but this universal information is made specific in each individual, becoming 'this one.' From an abstract point of view, all things are composed of atoms, but these atoms have different properties in different beings or things. The atom in my body is no different from that in the body of a pig, but the atom exhibits "pig nature" in the pig and human nature in me. In reality, nothing contains universality, which is only the product of abstract thinking. The so-called atoms with the same properties are just isolated atoms that exist as concepts, so to speak. In reality, no two existing atoms are the same. They are subject to the whole on which they depend and show the properties of the corresponding whole as components of the whole.

There is another problem with the above solution. One could say that if every process of the universe is holographic, then the information would be finite. In other words, each cycle shows the same information and then starts again after the display, definitely making the information finite. However, as mentioned above, one can argue that the universe is infinite, so its information should also be infinite. Regarding these questions, like the absolute inevitability of the universe mentioned earlier, we cannot give definitive answers, because these questions cannot be decided. This is because we are entering the realm of the infinite, and it is impossible for us to judge true from false (in finite terms) in this realm. This realm belongs to (philosophical) faith, and it can only be resolved by faith. It is impossible to answer the question from the perspective of finite knowledge, because human beings have no ability to understand infinity (in finite modes of cognition).

The above basic principles of holographic philosophy provide a new perspective for solving the problems raised by previous philosophies. At the same time, they also open up new fields of philosophical research. In the following two main sections, I would like to further elaborate on the holographic principle as well as on the correlation between holographic theory, reductionism, and systems theory—especially in terms of the whole–part relationship.

Further Explanation of the Holographic Principle

According to Norbert Wiener (1894–1964), information is neither energy nor substance. Definitely this does not mean that information can exist without substance and energy. In fact, it can exist only by relying on substance and energy. Information, substance, and energy are interdependent as a whole. They are only the expression of the same thing at different levels and in different aspects. In other words, they are different manifestations of the same reality.

Therefore, information can be defined as an attribute or mode of existence of matter, energy, and mind, and it is the material-spatio-temporal order of energy and mind. ('Order' here includes order and disorder. Information should not be understood only as the order of things, because disorder is also a kind of order, and therefore, it is also a kind of information. And information is the sum of order and disorder.) This means that information is the sum of all properties of matter. The term 'all properties' includes both known and unknown ones.

The most fundamental nature of information lies in its holography. The concept of holography originated from holographic photography. Later, Yingqing Zhang, a Chinese holographic biologist, transferred it to biology and discovered the law of biological holography. So can the holographic law be used as a philosophical concept to summarize the universal holographic phenomenon in the universe? The answer is yes. The boundaries of philosophy are fuzzy. There is no absolutely clear boundary between the concept of philosophy and the concepts of each specific science, and philosophy should not close itself off with an iron fence.

The holographic principle can be defined both in scientific and philosophical terms. The meaning of holography in this text is that the same information is contained in the parts (subsystems) and the same information is contained in these parts and the whole, and that all the parts contain all the information of the whole. As the basic proposition and principle of the theory of cosmic holographic unity, it reveals the wholeness and indivisibility of information as well as of the world, or the unity and identity of the universe in the sense of information.

It can be said that holographic theory represents a necessary consequence of the principle of universal interconnectedness in philosophy, and that it means the in depth-study of the latter. In terms of its manifestation(s), the universe means a universal interconnection, and all things are in an all-round interconnection, interaction, and mutual restriction. This makes every thing and every site of the universe to have all the information of interaction, interconnection, and mutual restriction from any point of the universe. Therefore, every thing is an information microcosm of the universe.

In this way, each subsystem "diffuses" its own information in all directions to act on all other subsystems and systems. Each subsystem is connected to all other subsystems and to the whole (infinite) system. At the same time, it also features internal cohesive relations. We call the cohesive relations of the subsystems holographic elements, and the whole system is the holographic system. All the information of the whole system is accumulated within the holographic element. Its explicit structure indicates the internalization or internal manifestation of its correlation and interaction with other subsystems and systems. In other words, its properties are only the confirmation and manifestation of the properties of all holographic elements and their relations.

Such internal and external relationships are essentially homogeneous and isomorphic. They differ only in form: extrinsic relations are latent and intrinsic relations are manifest. Let any holographic element be a holographic set Q. Each relationship has a corresponding element in Q. As holography makes this relationship of each thing infinite, the number of elements in Q are n… (infinite), thus constituting a holographic set (infinite set):

$$Q = \{n\ldots\}$$

Every thing and system is a holographic set like Q. It is through these relationships that each holographic element is in an all-encompassing exchange of information with all other things and systems as a whole. The above explanation of holography is only external and explicit. This refers only to external holography, but the fundamental reality is internal holography. Internal holography is of ontological concern. Specifically, things inherently contain the essential information of other things, which is the basis for one thing to be transformed into other things. Otherwise, the transformation would not be possible. The difference of things is only apparent because there is a difference in the manifestation of implicit information. In the latent state, however, all things contain the same information or attributes.

In this sense, there is no independent or immutable finite entity in the universe, and there are no absolute boundaries between all systems, because entities exist only in cohesive and interpenetrating relationships, and things are only an aggregation of relations, that is, a group of networks. Once these relations are cut off, the respective specific system will disintegrate. This represents the non-locality or globality of holographic elements. Holography is the root of the fuzziness of things, because holography shows that things can be this and that, or that and this, and they are contained within each other. Moreover, holographic theory shows that things are only an aggregation of relations that are always changing, uncertain, and comparatively fuzzy at all times.

Therefore, we can behold a wonderful cosmic hologram in which all systems, all things, and all levels of the universe endlessly permeate, contain, transform, cover, absorb, and reflect each other. The big universe contains the smaller universes, the big system contains the smaller systems, and the higher level contains countless lower levels. Since the latter are the epitomes of the former, we call this "Chinese box-like" structure a holographic recursive structure.

All of these depend on each other. One thing depends on all things, and all things also depend on every single thing. It can be said that one thing is composed of all things, and it also constitutes all things. All things in the universe interact with each other and are in dynamic holographic connection. The intrinsic information is exchanged at all times, forming a unified cosmic harmony.

One might ask, since every part of the universe contains all the information of the whole universe, and now it already contains the information of the future, then why do things still develop and unfold after all? Is that which will be new already inherent or new in the strict sense of the word? The new information emerges in the development, doesn't it? If there *are* things that *will* emerge in the future, can we say that the world is developing? In fact, those who ask these questions have only part of the universe in mind, rather than the whole. In other words, they only take into consideration the manifest universe.

The universe exists in two ways: the manifest universe and the potential one. Accordingly, information exists in two forms: manifest information and latent information. The world of latent information is infinite, while the world of manifest information is the limited expression of latent information.

Latent information is the information that is "hidden" in highly concentrated form that functions as a connector and has not yet been expressed through material or energetic carriers. Manifest information is in the opposite state. For example, the seed of a plant is a pure information node or holographic set of latent information. All the information of the future development of the whole plant lies in the seed without any exposure. Later the information displayed in the whole process of plant growth becomes manifest information. However, throughout the whole process of development, all the information contained in the original seed still lies in every part of the plant (all cells) all the time. The same is true of the entire universe. All the latent information in the "singularity" of the universe exists at every moment and at every point in the universe. Although it is infinitely latent and we can hardly find it, the same truth can be inferred from the fact that the basic particles originally formed in the universe still exist in the present universe and constitute the foundation of all things.

Therefore, the development process of all things does not indicate that things simply come from nothing, or that new things negate the existing ones. On the contrary, it indicates a full confirmation of latent information. In other words, *it is a process* in which latent information is transformed into manifest information under the active influence of external things. Only from this perspective can we understand how the development process of things can indicate both negation and affirmation at the same time. Everything is the sum of latent information and manifest information, and its development process is the manifestation process of latent information. In terms of the combination of these potential and actual states, this process means both the affirmation of latent information and the negation of manifested information. Each newly manifested information displaces the previously manifested stage and makes it latent.

From this point of view, the overall development process or manifestation process of things is characterized by an increase in order and the complexity of this order, such as the development process from inorganic matter to organic matter and finally to organic living beings, and the growth process from seeds to plants. In other words, the manifestation of things is becoming more and more complex. However, from the perspective of the unity of latent and manifest states, the amount of information has neither decreased nor increased, because the individual decline of things does not mean the disappearance of information, but only the process of transferring information from the manifest state to the potential state, that is, the latency process of information. If the development process of any thing is G, its manifestation process (or manifested information) is $+G$, and the latency process (or latent information) is $-G$, the following set results:

$$G = \{+G, -G\}$$

As the sum $G = 0$, the amount of information is neither increased nor decreased in the process.

Since every position and every thing is a holographic set, why can one receive only a limited amount of information? The reason is the hierarchical nature of information. The holographic expression of information is regulated, and the mechanism that controls the expression of information can be called holographic order. The manifest things are all in space–time, and the determined space–time is always limited. On this limited stage, only limited actions can be performed. This limitation is the requirement for the orderly expression of holographic order. What kind of drama would it be if all the actors would appear on stage at the same time? Information can only be presented layer by layer in a certain order. Therefore, controlled by the holographic sequential order, the latent information can only be performed sequentially on the stage in finite time and space, and only some information can be expressed, just as we can only think of a limited number of things in the instantaneous moments of time and space.

The holographic order specifies the length of the information band expressed by each thing, that is, the threshold of its manifestation. Once the information manifestation process of a thing reaches the specified threshold, it will "bounce" up (manifestation) or down (latency) to appear in another state. This threshold is called holographic threshold.

Due to the difference between the manifest state and the latent state, the degree of holographic convergence between things in the manifest state is different, which is called 'holographic degree.' The more similar things are, the higher the holographic degree is, and vice versa. Holographic degree refers only to the explicit expression of holographic order. From the perspective of the general unity of the manifest and latent states, the holographic degree is absolute, or rather, there is no difference in holographic degree.

In this sense, the information that makes things holographic does not come from the outside of things, but from the things themselves. All things contain the same information, which is the same as the information of the universe as a whole. However, the latent information can only be gradually manifested through the interaction and interconnection with other things and activities in the whole, and herein lies the significance of the interaction and interconnection of things.

Things at each level (the concept of 'level' herein includes temporal levels, i.e., stages) are the expression of particular levels or bands of the original information of the universe. The long-term and short-term sustainment of this specific information indicates the life of things, namely the phase during which a thing exhibits a certain stable pattern. After the expression of information at that level, the thing at that level will dissolve at that level and be transformed, that is, transit to other levels and states.

Here another question arises: How does all the information of the universe exist in each holographic element? Like in biological evolution, cosmic evolution is also realized through 'heredity.' Although the mechanism of cosmic evolution is not clear, there is no doubt that the most fundamental level of the universe is preserved in every "generation" and part, just as the original cellular form is preserved in every living being. Any higher-level thing must rely on the things on the lower levels generated levels, and it must encompass all those levels, without which evolution would not be possible. All things are homologous with the same foundation, and they are all built on the basis of the original universe. Therefore, in the transferred sense, it can be said that there are cosmic "genes" at the lowest level of all things, and all things are just the expansion of cosmic "genes."

In short, the universe cannot be created from nothing. The process of things evolving is predetermined, and the essence of this process is the expansion of the internal information or latent information of things. Although this expansion is inseparable from the interaction with external conditions, the external effect will not change the fundamental nature of things, but only the quantity of things.

Holographic Theory, Systems Theory, and Reductionism: A New Perspective for Understanding the Whole–Part Relationship

As already mentioned, philosophical holography is based on the idea that 'the part contains all the information of the whole.' This proposition provides a new perspective for understanding the relationship between the part and the whole, that is, the part has wholeness, and the whole also features partiality or locality. From the perspective of holographic theory, it can also be said that the part is the same as the whole, because each part of the whole contains all the information of the whole, and the part is a potential whole that has wholeness. In other words, each part is a small whole in transition toward the (absolute) whole. Systems theory and reductionism share common limitations and feature the nature of reduction. Systems theory is just the "inversion" of reductionism, since it attributes the nature of the part to the whole, and reductionism attributes the nature of the whole to the part. It is impossible to look at the problem as a whole, because the so-called whole is only the whole in the human view, and any human understanding is *not* 'complete.' Therefore, the so-called whole is essentially only a 'part' of the understanding. A complete understanding must include the 'unknown'; otherwise it is not 'omniscience.' Any known cognition is not omniscience as all of these represent limited, partial ("localized") knowledge. To use this incomplete understanding as a substitute for omniscience, which includes the unknown, means to replace the whole with the part.

New Interpretation of the Whole–Part Relationship

Historically, there have been two main representative views regarding the relationship between the part and the whole: that of reductionism and that of systems theory. Reductionism attributes the nature of the whole to the nature of parts or lower levels. Systems theory holds that the whole is greater than the sum of the parts and that the nature of the whole cannot be attributed to the parts. The perspective of holographic philosophy provides a new understanding of wholeness. According to holographic philosophy, wholeness is complex, and some single relationship can hardly capture this complexity.

From the point of view of holographic theory, if the relationship between the part and the whole is understood as 'greater than' and 'lesser than,' there are at least three aspects to describe their relationship:

First of all, the whole is greater than the parts or the sum of its parts. Every proposition has its premise, and it is not true in every sense. The premise of this proposition is that the whole is greater than each part in quality and quantity, which is obvious and understandable from the perspective of the sensory level. Therefore, if the proposition is based on this, it has no theoretical value. This is the meaning of 'the whole being greater than the part' in a sense of a "mechanical sum" criticized by systems theory. The proposition that 'the whole is greater than the sum of parts' of systems theory is valid only from the perspective of function, that is, the function of the whole is always greater than and more comprehensive than each part, as well as also greater than the mechanical sum of each of those parts. Alternatively, some functions of the whole are always not possessed by its parts. The viewpoint of systems theory is valid only from this perspective.

Without the two perspectives, the proposition that 'the whole is greater than the sum of parts' would be difficult to establish. This can be demonstrated in view of the following two aspects:

From the perspective of development and evolution, a part can also be greater than a whole. No part is passively constrained and determined by a whole to always remain in the original state. On the contrary, it always tries to break away from the whole and to become an independent individual, that is, a new whole, and this is what is called development and evolution. As the process develops, the original part will far exceed the original whole in quantity and quality, not to mention in vitality, and the original whole will gradually lose its vitality and even disappear in the process. For example, a tree can give birth to countless new trees, and a person can have many children. Since these are all obvious facts, they have little theoretical value. They are listed here only to illustrate the complexity and diversity of the relationship between whole and part, and that there is more than one relationship between them to consider here.

From the angle of information, the part is equal to the whole, that is, the part also has the wholeness of the whole. Wholeness means that the part also has all the information of the whole, and it is also a small whole in itself. Although a finger cannot have the function of the stomach, it contains the information of the stomach.

It should be noted that the part contains the information as a whole, not the actual whole or the functional whole but the wholeness of information. Each part of the information is the information as a whole.

Since the wholeness of a part is not valid in a functional way, in what way does it exist? As mentioned before, it exists in a latent or undiscovered way, or it can also be called latent information. From the point of view of things themselves, the latent information is the virtual information that has not yet become manifest. For example, the information in seeds is latent information in relation to trees. From the perspective of human beings, latent information also refers to the information that people have not yet been able to cognize.

Some people may wonder how come the latent information can be known. And others might even argue that latent things are not the subject of study by science. This is a strange point of view. Does science only study things that are visible? Then what is the point of studying such things? After all, is not science always about exploring the unknown, the latent information, because the latent is the unknown in terms of human understanding? To explore the unknown is to reveal the latent information. Moreover, it is necessary to demonstrate how to know the existence of latent information. Is there any information that has not been manifested in a stone? The nature of 'stone' is just a kind of information that has been manifested. What you see is just a "seed," but in the "seed" there are countless aspects of information in the seed that have not (yet) been unfolded, and that information is latent information. The development and evolution of things is the process of manifesting information. If all the information would be manifested, things would not develop and evolve, because there would simply be no information to manifest. As long as things are still evolving, there is still information that has not yet been actualized.

Therefore, what holographic theory reveals is the maximum wholeness. In other words, it indicates that not only the whole itself has wholeness, but also all its parts have wholeness. It can be said that this is the most complete wholeness. The commonly understood wholeness emphasizes the dependence of the part on the whole or the constraint that the whole imposes on its parts, while the wholeness in holographic theory refers to a total wholeness, that is, in a whole, not only the whole has wholeness, but also each part has wholeness as well. This means that an absolute wholeness or unity is maintained in every whole. This (absolute) unity or wholeness is an informational unity. What circulates between the parts and the whole is information that is mutual, and the transmission of the information has carriers that can be described in terms of energy, material, etc.

This kind of wholeness indicates that the parts are not predetermined by the whole and that each part has its autonomy that comes from its wholeness. Autonomy means that each part has the information and inherent power to become a whole, to try to become a self-determining whole.

So why can a small part become a new whole? Why does it have this potential? Why can the part have the power to move forward and to evolve? It all lies in its wholeness and in the wholeness of information. Wholeness empowers it to expand and to evolve. Since the part implies wholeness, it tries to be complete. Only a complete thing has autonomy, independence, and the power to expand. In other words, only the whole has power. Since the small part implies a whole, that whole has to manifest; otherwise it would not be a whole.

If the part were to contain only fragments of some information or fragments of the whole without any wholeness in itself, it could not have such power. If it had no inherent wholeness, it would have no goal. What can a fragment become? It has nothing to become. Since it is already a fragment, it has no need to evolve. This explains why it does not have the power to evolve. Even if such a fragment were to have the "intention" to become a whole, it would not have that power. How can a single fragment generate the power to push itself forward? It can only be pushed, but it has no ability to push independently. The part that contains the wholeness manifests as an organic being due to the existence of this wholeness. All that is organic comes from wholeness. Only as an organic whole can it produce the power of self-evolution and independent development. This is perhaps what is often called 'organization.' Organization comes from wholeness, or organization is wholeness. Without wholeness, there is no capacity for organization, and things without the capacity for organization can only be fragments.

In this way, the latent information is not passive, nor is it permanently confined as the latent in nature. On the contrary, it always penetrates the "provisions" of the whole for itself, becomes manifest, and then evolves from the status of the part to that of a (more evolved) whole.

Every thing and every part is an infinite whole—a part of the universe. Therefore, every thing and every part contains infinite information. But this infinite information can only be partially revealed in finite space and time, while all the other information remains hidden. Therefore, everything becomes an ocean of infinite latent information, and the information it manifests is only a little spray from this ocean of (infinite) potentiality. At the same time, every part or thing is a potential whole in the process of tending to become a higher, more evolved whole.

Since all parts have wholeness, what are the differences between the whole and the part? There are still some differences. A part is a latent whole, while a whole is a manifest part. The difference between the two is in the state of existence, but there is no substantial difference in information. Except for the whole universe, every whole is not always just there, but is formed gradually. At the beginning of its formation, it is part of a "parent body." *Therefore, the whole is only an evolutionary stage of the part, just as the part is also an evolutionary stage of the whole.*

It can be seen that the part and the whole are not isolated. They are different states of the same (processual) existence. Even as specific parts and wholes, their nature is not separate but *mutually constitutive*. It is only in such a whole that the part has such a nature and contains such a wholeness. Similarly, since there are such parts, there will be such wholes, and the wholes will have such a state of existence. Strictly speaking, it is meaningless to discuss the relationship between the whole and the part in the sense of who establishes whom. If we can say anything at all here, we can say that wholes and parts just regulate and establish each other.

With this relationship of part and whole in mind, we cannot just say that the whole is composed of parts, which is often emphasized as being obvious. We should also say that the part is also composed of the whole. The part is actually the manifestation of the whole in this particular region and time. It contains all the information of the whole, the latter of which is implicit in the part as a constitutive entity of the part. The whole constitutes a part, just as the part constitutes the whole. Since everything is a part of the universe, it can be said that everything is composed of the whole universe.

Thus, the boundary between the whole and the part disappears, at least from the perspective of existence itself. In terms of existence itself, there is no distinct boundary between part and whole. In reality, in which thing can we find the boundary between whole and part? We cannot find it. Existence itself is neither 'parts' nor 'wholes,' and everything just exists like that. The division of part and whole is mainly drawn by human observers. We draw a line in the whole and say this is the whole and this is the part. The original internal relationship between the whole and part is then "cut off." It should be said that the division of the whole and the part is mainly for the convenience of cognition and for the need of expression. Without such a "line," we cannot describe the state of existence of things. Therefore, from our human perspective, the real meaning of the whole and part is actually in the different perspectives. From one point of view, it is the whole, and from another point of view, it is the part. Different points of view will also produce different results. For example, from the perspective of systems theory something is a part, but from the perspective of holographic theory it is a whole. In this sense, 'part' and 'whole' are just two terms that we have invented and developed.

From the perspective of things themselves, a part must be a part of the whole, and the whole must also be the whole of its parts. The whole without parts and the parts without a whole cannot be called 'part' or 'whole.' It is the whole that defines the part as a part, and it is the part that defines the whole as a whole.

Holographic Theory and Systems Theory

On the surface, the viewpoints of holographic theory and systems theory seem to be mutually exclusive. According to holographic theory, the part is a whole. According to systems theory, a whole is greater than the sum of parts. However, in essence, the two are consistent, or in a sense, the holographic theory is the inevitable consequence of the further deduction of the principle of systems.

According to the point of view of systems theory, the parts of the system are specified by the whole. Therefore, it is inevitable that when the whole system specifies its various parts, it will input the information of the whole itself into each part. The parts are specified by the whole in terms of matter and energy as well as information. At the same time, in the process of interaction between whole and part, the information of each part is naturally fed back to the level of the whole system. Thus, the information mapping and correspondence between the whole system and each part are formed, and each part contains the information of the whole system, resulting in the holographic relationship between them. It can also be said that the holographic theory is an inevitable conclusion of systems theory. In other words, a complete systems theory must be holographic theory, and the further development of systems theory leads to holographic theory.

Some people judge the holographic principle from the standpoint of systems theory and think that the two are opposed. This view represents the narrow view of traditional monism (which, in my opinion, is actually a kind of dogmatism). From the perspective of holographic theory, systems theory does not contradict holographic theory. Their difference lies in the respective perspectives. Systems theory is about operation and function, and holographic theory is about information. From different angles, we will naturally see things differently. One cannot say which is right or wrong because different perspectives have their own intrinsic validity and cannot replace each other. In addition, and as just mentioned, it has to be emphasized that in terms of the content of the theory itself, the further development of systems theory leads to the perspective of holographic theory—although, in fact, holographic theory is not derived from systems theory. In other words, the holographic relationship between part and whole is an inevitable conclusion of systems theory, while, on the other hand, the viewpoint of holography cannot replace that of systems theory, just as the viewpoint of systems theory cannot replace that of holography. There is no need to argue about that any further.

Another point is that, according to systems theory, we should see everything from the perspective of the whole, but in fact this is impossible. One reason is that since we are talking about "seeing" (in the transferred sense), things will be "seen" from a particular angle. Therefore, the things we reflect upon will not be represented in their entirety. If we want to "see" the whole, we have to take it into consideration from all angles; otherwise we will not see the whole, but the parts. Moreover, even if things could be "seen" from all angles, this would only refer to the cognition that one already has, and our cognition is always limited. Therefore, no matter how "complete" or "holistic" one's knowledge seems to be, it will still only be nothing more than the knowledge of some of the properties of the object (of possible knowledge), that is, it is still limited cognition. Thus, (our) cognition can neither represent the whole nature of an object nor the wholeness of things (as such).

Another reason is that each whole has an infinite nature, otherwise it would not be a whole. The whole must be complete, and incomplete things cannot be whole. In this way, no matter how small a whole is in its wholeness, we cannot find its limit, even in view of a grain of sand. To identify the wholeness of a grain of sand, it is necessary to understand how this grain of sand came into being. However, we will not be able find its ultimate origin by looking backwards, because we would have to trace it back to the formation of the universe itself. "Looking forward," how can we know what this grain of sand will evolve into? We cannot trace the final destination of its developmental path. From a macroscopic point of view, in order to understand all the properties of this grain of sand, we would have to be able to completely understand its relatedness to the universe at large. At the smallest level, we cannot even trace its micro boundary.

The conclusion here can only be that what we call the whole is always only a part, a part of one's recognition and understanding of things.

Therefore, it is basically impossible to see the problem as a whole. What is often referred to as 'seeing a problem as a whole' is actually the perspective of the so-called "whole" as *someone* understands it. If we look at a problem from such a "holistic" perspective, we only look at it from the perspective of someone. In essence, this still means to look at a problem from a certain angle (i.e., in the sense of a part) that fails to represent the whole.

For the above reasons, holographic theory comes to an epistemological conclusion that differs from conventional epistemological points of view. In the sense of holographic theory, a complete knowledge must include the unknown within itself, otherwise it could not be considered 'complete.' The unknown is a fundamental component of knowledge. This is not to be understood in the sense that the unknown is a prerequisite for knowledge. Rather, the unknown is itself a component of any knowledge. If we admit that the world is infinite, we will come to this conclusion. If the world is infinite, then the nature of every thing in the world is infinite, because it can have an infinite way of existing in the infinite world, and because it has an infinite relationship with the infinite realm of things. We simply cannot understand all of its properties. As Nicolaus de Cusa (1401–1464), a German Renaissance philosopher, said, since the world is infinite, the parts we know are always limited; since we do not know the infinite world, we do not know the nature of the known world at all, because we do not know the relationship between this known world and the unknown world, and we do not know the meaning of the known world in the unknown world. Therefore, we do not know the fundamental meaning of the known world in the sense of the unknown world. In this case, everything we know contains the unknown and our non-knowing. If we exclude the unknown from the known, we arrive at a very superficial form and understanding of knowledge, or rather, incomplete knowledge.

A very characteristic example to illustrate this problem can be derived from the context of traditional Chinese medicine (TCM). At present, there are three main production methods of TCM: decoction, pills, and Western-style TCM (such as orally administered liquid, the modern type of tablets, etc.). Based on the author's experience and research, the effects of these three types of traditional Chinese medicine are very different. Their curative efficacy is in the above-stated order. Decoction is the best, traditional pills are the second-most effective, and "Westernized" forms of TCM are the least effective. For instance, in the case of the herbal medicine *shengmaiyin* 生脉饮, administering it in the form of a decoction has a good curative effect, but the Western-style modern type of ready-made liquid has no effect. Why is there such a big difference? It is because the production method of Western-style medicine only represents knowledge without the unknown. The product is extracted and processed by analyzing various components and elements in TCM. In other words, it is made according to the existing knowledge of mankind. After being refined, the elements that are considered useful in Western-style medicine are retained, and those that are useless or seem to be useless are discarded. However, after such a process of purification, the unknown and the unknowable are "refined away." Humanity's current knowledge only recognizes a certain level and only limited aspects of things—certainly not the whole of things. The elements extracted by man mean only some parts or aspects of these things and not everything that these things are. That is why their curative effect is naturally poor, or why they are even ineffective. In TCM, the reason why the administration of medicine by way of decoction is so effective is that it contains the unknown. We only know from experience that some drugs will have certain therapeutic effects when combined, but we do not know how these drugs work, what factors are at work in the whole, and what relationship and function between these drugs leads to the corresponding therapeutic effects in toto. The elements we know are only a few, and they are not enough to summarize all the relations between them and their properties. *The effect of traditional Chinese medicine when administered in the form of a decoction is actually produced by the combined action of what we know and these unknown aspects and factors.* Western-style extraction and "refinement" negates or ignores these unknown factors. In this way, what is refined represents only some parts of the original whole, and the therapeutic effect is naturally poor. By relying on the method of decoction, TCM (implicitly) realized an omniscient method, that is, it includes the unknown. This omniscient method contains 'non-knowing' in itself. Here we see that the unknown is not useless, but of great eminence. Genuine knowledge, or supreme knowledge, must include the unknown. Otherwise, it cannot be considered as genuine or omniscient knowledge. How can an incomplete knowledge be genuine knowledge? As Chuang Tzu (Zhuangzi 庄子, c.3rd century BCE) said, knowledge that contains the unknown is genuine knowledge.

From this, we can draw the following further conclusions. First of all, from a finite human perspective, the whole as such must contain the unknown, because the whole is infinite. It is infinite in time, space, quality, and quantity. On the other hand, what is often referred to as "whole" often only refers to known parts and it is essentially limited. This means that what we call a whole in the latter sense is still only a part, it does not represent the whole of things. Only the reflection of a whole that includes the 'unknown' is truly referred to as a whole, or only that which includes the unknown is a whole. Secondly, all possible knowledge is imperfect, because it contains the unknown, and there is an infinite field of unknown aspects behind everything that is known. Thus, if any (explicit) knowledge is considered complete, it will inevitably lead to a very narrow view.

It can be seen from the points outlined above that systems theory, including related forms of "holism," is essentially an inverted reductionism that attributes the nature of all elements in the system to the whole, and that this "whole" only represents the actuality of limited human cognition. It makes a similar mistake as reductionism, which simply attributes the nature of things to a particular level—but only in the opposite direction.

Holographic Theory and Reductionism

At first glance, the holographic principle may seem no different from reductionism, which holds that the part is the whole. However, according to the further implications of both points of view, there is a substantial difference.

Holographic theory understands the wholeness contained in the part in terms of the information contained in the part, that is, the identity of part and whole lies in their identity of information. It is the identity of the information contained in the object itself, rather than an identity of actual existence at a particular level, that constitutes the wholeness here.

The identity between part and whole studied by reductionism is to be understood in the sense of actual existence, that is, it always reduces the nature of the whole to a particular level or part of the whole. For example, it reduces the wholeness of something to the atomic level or to chemical movement, and refers to this particular level and movement to explain the nature of the whole. For example, it is argued that the world is made up of atoms, and therefore everything functions in various ways as an expression of atomic existence. The essence of reductionism is to dissolve the individual and particular into the universal, or to universalize what was originally particular. Although this view reveals the sameness of things in terms of material aspects as well as a kind of universality of composition, it also has its limitations. It cannot explain the origin of individuality and particularity and why there are such great differences between different things made of the same sort of atoms. For example, it can not explain why pigs and humans are greatly different although both are composed of atoms. This means, reductionism fails to explain the source of the characteristic information of the whole itself.

What holographic theory reveals is that any part and every level contain *all* the complexity of the whole, and this complexity is the source of individuality. Since this complexity contains the entire history of each being itself and all of its relations with other things, each thing exists in a special, unique space–time structure, so that everything is different from everything else and cannot be replaced by anything else.

Although all things share holographic relationships and contain the same information, this same information is uniquely specialized or "individualized" in the manifestation of each thing, making the particular unique "stone-ness" in a stone and the individual expression of a "pig-nature" in a pig. A part holds all the information of the whole, which means that parts contain all the complexity of the whole, that is, a part contains all the differences. Difference is information. Without difference, there is no information. The formation of every thing and even every activity of things are cosmic events and expressions of all the laws of the universe. Without any of the laws of the universe, things could not be generated and their activities would not be possible as they are. From this point of view, it seems that every thing and every activity is a universal activity that symbolically expresses universal laws. In fact, this is not the case. In every thing, at this moment, the whole of all the seemingly universal regularity is in fact individualized in unique ways in each case, and the most universal thing becomes the most special aspect in the individual. In other words, the most universal aspects exist or are expressed always in a most particular way, and all the information is reconstructed in the most special and individualized direction in the individual, making it a unique "this one." The nature of the same kind of atom in one individual atom is different from that in another, because it is in a different space–time or environment, and it contains all its space–time structure within itself. In individual things, universality and particularity are integrated. Individuality contains all universality, and universality becomes (manifest) particularity. The existence of particularity is based on the wholeness of universality. There is nothing that is not a particular, just as there is nothing that is not universal (at the same time). The most universal thing exists in various specific occasions. There is no universality separated from individuality—any such "universality" can only be a human abstraction and does not exist in reality.

The problem here is that these so-called universal factors in any individual take on the nature of that individual, because the universal factors are subordinate to the individual nature in the individual, and, at their core, they are constituted in correlation with the nature of the level at which the individual is located. For example, in pigs, atoms are subordinate to the nature of pigs, and atoms are embedded in the form of the nature of the whole pig, so they exhibit the nature of pigs. In a particular stone, the atoms exhibit the nature of the stone because they incorporate the nature of the stone as the center of their construction. In other words, in an individual, the information that appears at the level it has reached in its evolution takes a dominant position, and all the information, matter, and energy it has can only be subordinate to the dominant information. Therefore, all the matter that constitutes an individual takes on the nature of that individual.

According to the reductionist interpretation, both humans and pigs are made up of atoms. But it is impossible to distinguish humans from pigs at the atomic level. It is difficult to explain how things made of the same kinds of atoms can be so different. What can be said is that these differences are certainly due to the overall conditions. However, the respective information (properties) does not come from the outside, but from the thing itself. In a certain sense, it can be said that it also comes from the atoms and any other level of the thing, because they all contain information of the thing as a whole. As for why the same atom shows such different properties in pigs and humans, it is due to the provisions and influence of the whole (it should be noted that this whole consists of all the factors of the whole space–time related to the thing, including the level of the atom), which also determines the direction, degree, and mode of information manifestation of atoms in different individuals. In this sense, it can be said that atoms also have "individuality." The atom of a pig has a different nature from the atom of a human being. The properties of atoms in a particular thing or being are different from the nature of atoms in all other things or beings, and the generalized atom is just a product of human abstraction.

To summarize this final point, holographic theory or holographic philosophy overcomes two tendencies. The first is reductionism in the traditional sense, which holds that the whole can be reduced to its most basic level, such as reducing things to atoms or more basic particles, and reducing the life process to physical and chemical processes. The other is the previously mentioned form of "holism," which links the nature of the part to the whole. This view is essentially the opposite of reductionism. Both of these views share a common limitation, that is, they antagonize the part against the whole, thus necessitating a decision as to which side determines the other. In fact, the state of existence of a system or a thing is not determined by the whole or the part, but by the inseparable relationship, reciprocal integration, and interaction between the whole and the part, each of which can only realize itself in the other.

Notes

1. In this text, the transliterated names of modern Chinese authors are written in the "Western" order of given names followed by family names. In the reference list, the transliterated names of the Chinese authors are also presented with the initials of the given name followed by the family name. Since given names that consist of more than one Chinese character are transliterated into Pinyin as a single unit, only the first letter of the transliteration of the first character of the given name is used as the initial here. In Chinese characters, the original order of family name followed by given name is maintained, of course.

References

O. Spengler, 奥斯瓦尔德·斯宾格勒,《西方的没落》[*The Decline of the West*]. 陈晓林 (译) (黑龙江教育出版社, 哈尔滨, 1988)

C. Wang, C. Yan, 王存臻、严春友, 宇宙全息论 [The holographic theory of the universe]. 自然信息 (2/1), 1–5 (1985)

C. Wang, C. Yan, 王存臻、严春友, 精神全息论 [The holographic theory of mind]. 现代哲学 (4), 27–28 (1986)

C. Wang, C. Yan, 王存臻、严春友, 宇宙全息重演律 [The cosmic law of holographic reiteration]. 福建论坛 (2), 74–77 (1987a)

C. Wang, C. Yan, 王存臻、严春友, 精神全息重演律 [The mental law of holographic reiteration]. 北京师范大学学报 (3), 96–103 (1987b)

C. Wang, C. Yan, 王存臻、严春友, 宇宙全息统一论 [The theory of cosmic holographic unity]. 自然信息 (6), 25–28 (1987c)

C. Wang, C. Yan, 王存臻、严春友,《宇宙全息统一论》[*The Theory of Cosmic Holographic Unity*] (山东人民出版社, 济南, 1988a)

C. Wang, C. Yan, 王存臻、严春友,《宇宙统一科学》[*The Science of Cosmic Unity*] (山东人民出版社, 济南, 1988b)

Z. Wang, 王兆强 , 广义全息与全息不全——兼与《宇宙全息统一论》作者商榷 [Generalized holography and holographic incompleteness: a discussion with the author of *The Theory of Cosmic Holographic Unity*]. 科学技术与辩证法 (6), 5–9 (1989)

C. Yan, C. Yan, 严春宝、严春友,《文化全息论》[*The Theory of Cultural Holography*] (山东人民出版社, 济南, 1991)

C. Yan, 严春友,《精神之谜——全息精神学》[*The Mystery of Mind: Holographic Noology*] (中国社会科学出版社, 北京, 1991)

C. Yan, 严春友, 全息哲学论纲 [Outline of holographic philosophy]. 华中师范大学学报 (增刊), 1–10 (1996)

C. Yan, 严春友, 全息论对系统论与还原论的超越 [The advantage of holographic theory over systems theory and reductionism]. 河北学刊 (6), 60–62 (2008a)

C. Yan, 严春友, 全息哲学引论 [Introduction to holographic philosophy]. 社会科学论坛 (12), 31–47 (2008b)

Chunyou Yan (China) is a leading expert in contemporary Chinese natural philosophy. His works combine various traditional Chinese views with Western philosophical elements and modern holographic cosmology. He is a professor at the College of Education for the Future and at the School of Philosophy at Beijing Normal University in China.

The Human Body as the Singing Universe

Bei Peng

Abstract For millennia, the basic idea that there is a universal order that connects human beings and the universe has lived on in many cultures. This order has often been expressed in geometric or musical-harmonic terms. From Pythagoras to Kepler, universal scholars were firmly convinced that this order represented the primordial code of all things. This chapter explores a new interdisciplinary perspective that combines the fields of traditional Chinese medicine (TCM), music theory, and Keplerian astronomical insights. By means of corresponding mathematical analyses, it is shown that the energy meridians of the human body correspond to a specific musical-harmonic order, which in turn is inextricably linked to certain relational aspects of the celestial bodies of our solar system. The study represents completely new discoveries in relation to this universal connection. In doing so, it also opens up new levels of understanding and provides new aspects for the further development of the concept of cosmic life.

Keywords Cosmic harmony · Acupuncture points · Planetary movement · Music therapy · Music theory

Introduction

> Every point in the universe is a miniature of nature, but in each of them
>
> the artist has copied the original from a different side.
>
> Johann Wilhelm Ritter
>
> Heaven and Earth live with me. The ten thousand things are one with me.
>
> Zhuangzi

B. Peng (✉)
Institute of Advanced Studies in Humanities and Social Sciences, Research Institute for Globalization and Cultural Development Strategies, Beijing Normal University at Zhuhai, Zhuhai 519087, China
e-mail: bei.peng@bnu.edu.cn

D. Bartosch et al. (eds.), *Towards a Philosophy of Cosmic Life*,
https://doi.org/10.1007/978-981-99-2131-7_6

Throughout human history, the concept of man as a cosmic being has been explored and established through religious and philosophical traditions in various civilizations. The concept of man as a cosmic being has always been part of human identity. This identity also reminds us that despite our individuality, we are all part of something much bigger: a universal, cosmic energy and order that binds us all together. As a species, we have a shared responsibility to work for a better world and to be aware of our cosmic interconnectedness. In this way, we can all strive to become more and more like the perfect human being that we are within. The understanding of the human being as a cosmic being is based on the important logic that we are endowed by nature with the harmonious order of the universe and that we can perceive this order both sensually and rationally. From this, different cultures have derived codes of conduct for human society to conform to this harmony. The order of the universe thus becomes the basic matrix of our human understanding of the unity of macrocosm and microcosm, spirit and matter, as well as the celestial movements and human society.

Myths and legends have existed since the dawn of human civilization. From the ancient Egyptians to the Norse gods, from the Aztecs to the Chinese legends, the creators of ancient civilizations have sought to pay the highest tribute to the origin of life and cosmic order in a myriad of symbolic forms. Hans Kayser, in his summary of the mythological narratives of various ancient civilizations such as ancient Egypt, Babylon, ancient China, and ancient India, points out striking similarities between the creation myths, all of which allude to a direct connection between the creation of the world and life and "sound" (Kayser 1950, XIII–XVI). This image of a primordial, mystical and otherworldly sound relates to a very ancient tradition of observing and studying acoustic phenomena in nature and to a reflection on the connection between the origin of life and celestial phenomena.

The great intellectual power of mathematical thinking has provided the keys to understanding the universe and the order of life since ancient times. The Pythagoreans recognized the close connection between mathematics and the physical world and used arithmetic and geometry to explain the world qualitatively and quantitatively, explaining the fundamental harmonic order of the universe through mathematical patterns. They also noted that music allowed people to know the same cosmic order in an audible and direct way, for "all heaven is harmony and number" (Aristoteles 1989, 31 [986a], tr. BP). At the same time, the sages of China believed that the five tones represent the five planets and that the harmony and unity between the tones is directly related to the planets. The absolute measure of the length of a particular pipe of the fundamental tone, the Yellow Bell, was used throughout the Chinese empire as the basic standard for deriving all other units of measurement, and it also served as the main tool for establishing social standards (even as late as the early 20th century) (Peng 2019, 138–142).

This holistic way of thinking that runs through the intellectual histories of both Western and Eastern cultures as well as the history of technology and art is based on the assumption that all parts of the universe are inseparable, that each part is part of a larger whole, and that each part represents an individual reality of that universal whole. Against this background, this essay attempts to analyze the relationship between the rules of universal harmony, the universe, and the human body from a new perspective. Within the framework of this book's focus on the development of a philosophical concept of 'cosmic life,' the present chapter seeks to formulate a new perspective from the interdisciplinary point of view of music theory, traditional Chinese medicine, and Kepler's theory of celestial harmony. This perspective is based on some phenomena that I have observed during my research in a cross-cultural context. Since I am a musicologist, the starting point for my research into this transdisciplinary topic was music theory. It is important to emphasize that music theory is not directly related to music practice in the usual sense of studying the harmonic, melodic, rhythmic, and stylistic elements of particular musical compositions, or musical styles, and so on. Rather, we are talking about a fundamental, universal system of music (in a derived sense of the ancient Greek concept of *theoria* [θεωρία]). Music theory involves a scientific observation of music as a physical-acoustic phenomenon and some principled mathematical calculations and representations derived from it. Since the field of music theory and its history have been sufficiently and extensively investigated, I will not discuss it here, except to provide some clarification whenever necessary.

Another important starting point for the present chapter is the theory of acupuncture points in traditional Chinese medicine (TCM). This system of medicine is widely practiced in East Asia and throughout the world nowadays, and it represents a knowledge of the acupuncture points of the human body based on a long history of therapeutic practice. The knowledge of the location of these acupuncture points has long been strictly prescribed and used for therapeutic purposes, supported by specific terminology. This chapter introduces an enhanced perspective. It includes the new hypotheses that (1) there are implicit numerical or proportional relationships between acupuncture points and that (2) these mathematical relationships are analogous to the numerical values of intervals in music theory. *This raises the question of whether the principles of musical harmony are also related to the principles of harmony for human beings as healthy living beings. This results in a new perspective for understanding the principle of the constitution and internal self-regulation of human life energy.*

The third angle of this chapter is related to Johannes Kepler's theory of cosmic harmony of the solar system. Kepler argued that the motions of the planets in the solar system are not random, but represent a musical harmony, and that the principle of this musical harmony is identical with the principle of the universe as it was known at the time. In the work of present-day researcher Hartmut Warm, Kepler's theory of the harmony of planetary movements is confirmed by new data. Drawing on his findings, I derive an analogy between the geometric relationships of the body's acupuncture points and the geometric relationships of the planetary motions in Kepler's world harmonics. This raises another new question, namely whether this principle of musical-harmonic order of the human body is also directly "compatible" with the principle of cosmic harmony. The consideration of such a hidden solar systemic "grammar" of life could open up a new understanding of 'cosmic life.'

Musical Harmony in the Human Body

In traditional Chinese medical theory, the idea of the human life energy system is central. Within this system of knowledge, the human body itself is viewed as a self-contained microcosm. In Daoism, the human body is referred to as a 'small Heaven and Earth' (*xiao tiandi* 小天地) coexisting with the 'great Heaven and Earth' (*da tiandi* 大天地), which means that "Heaven and Earth, space and time form one body with the human being" (*tiandi yuzhou yi ren zhi shen ye* 天地宇宙一人之身也) (*Huainanzi*, n.d., "Benjingxun" [§] 3, tr. BP). "All the orifices and limbs of man are connected to Heaven" (*kong qiao zhi ti jie tong yu tian* 孔竅肢體皆通於天) ([§] 21, tr. BP). This theory of the unity of Heaven and man assumes that human life as well as Heaven and Earth have the same origin, follow the same laws, and even form the same processual structure; and it is therefore assumed that they correspond and harmonize with each other on an energetic level. "The human being, as a phenomenon, unfolds into a multiplicity of individuals, in each of which the central One is included as the life principle" (Wilhelm 1929, 12–13, tr. BP).

What supports the effective implementation of *dao* 道 in the sense of this life principle in the microcosm and macrocosm is called *qi* 氣. This life energy of *qi* flows through the human body in an orderly fashion along certain biological energy pathways called 'meridians' (*jingluo* 經絡). From the perspective of TCM, these meridians can be understood as channels of a person's life energy in which *qi* is continuously circulated and transmitted as long as the person is alive—namely, in order to maintain and regulate the basic functions of the internal organs that constitute the body as an organic whole. Along the meridians there are small spots where *qi* energy concentrates. They are called acupuncture points. This ancient theory is also of great interest to researchers in modern medicine and biology. Scientists have been working on how to prove and measure the existence of acupuncture points in the context of modern medicine. They have produced some evidence for it. For example, in recent years, some scientists have discovered that bioelectric currents are active in the cytoplasmic matrix of connective tissue throughout the body of a living human being, and that stimulation of these acupuncture points, as mentioned in Chinese medicine, can have anti-inflammatory or pro-inflammatory effects (Langevin and Yandow 2002; Ahn et al. 2005; Langevin et al. 2007; Langevin 2014; Liu et al. 2020, 2021; Ulloa 2021). This proves the relationship of acupuncture points and meridians to connective tissue levels. The results of these modern medical experiments are reconcilable with the traditional theoretical system of Chinese medicine.

I would like to present an original and new view that is consistent with the current theory of acupuncture points in the modern context of the application of traditional Chinese medicine: The spatial arrangement of acupuncture points along the meridians on the surface layers of the human body is not a matter of chance and random arrangement. There is also much more to this new perspective than the rather elementary and widely accepted view, according to which the meridians are organized symmetrically around a central axis dividing the body into left and right sides and "mirror images" of each of the twelve main meridians on either sides.

In my view, the distribution of these acupuncture points is organized in complex proportional relationships to form a much more complex geometric structure. If we consider these proportions and geometric structures in the context of music and harmony theory, it becomes clear that the distances between these acupuncture points correspond exactly to the mathematical proportions of musical intervals, which means that the distribution of acupuncture points shows the regularity and harmony of the body's energy from a certain hitherto unrecognized dimension. *From this point of view, the human body, in its energetic dimension, can be viewed in the transferred sense of a delicate "musical instrument" that plays the song of life.*

For those readers who are not yet familiar with the theory of traditional Chinese medicine, I would like to begin with a brief introduction of the basics of acupuncture points. It is assumed in TCM today that, theoretically, there are approximately 720 acupuncture points in the surrounding tissues of the human body (under the surface of the skin).[1] The theoretical knowledge of acupuncture points in Chinese medicine has been gradually accumulated through millennia of experience, research, and practice. The Chinese term for 'acupuncture point' is *shuxue* 腧穴, which is sometimes alternatively translated as 'acupoints' (for more on the translation of this term, see also Xu and Zhang [2005]). In the Western context of applied TCM, it is also referred to in Latin as a *foramen* (Hempen 2009, 67). The word *shu* 腧 in Chinese means an energetic "gate" of the living body, and *xue* 穴 similarly means an opening, a cavity, a depression, an access, or an "entrance." *Shuxue* are the "points" on the surface level of the body where *qi* flows from the meridians and from the internal organs to "enter and exit," to converge, and to be transmitted. These energy zones, which are theorized as points in the sense of small spots on the surface of the body, are divided into three categories: regular, extra-meridian, and *ashi* 阿是 acupuncture points (Liu and Liu 2009, 11–12). They are located in different areas of the human body and form a stable energy network. According to traditional Chinese medicine, these acupuncture points can be stimulated by means of various therapeutic agents to prevent or treat diseases.

A practical method for identifying acupuncture points that has been developed on the basis of thousands of years of experience is called *shuxue dingwei fa* 腧穴定位法, "acupuncture point localization" (Sui and Meng 2013, 1–4). According to this, each acupuncture point has a name and a fixed location. The purpose of this teaching is to mark the precise locations of acupuncture points on the surface of the body, so that the physician can learn to accurately press or hit them with an acupuncture needle. In the earliest medical compendium, *Huangdi Neijing* 黃帝内經 (2nd century AD), several chapters detail 160 major acupuncture points and their therapeutic effects (Unschuld 2003). During the 16th and 17th centuries, there was also a great deal of research into earlier developments in Chinese medicine. In 1601, the Chinese physician Yang Jizhou 杨继洲 (1522–1620) published *The Complete Book of Acupuncture and Moxibustion* (*Zhenjiu Dacheng* 針灸大成)—a ten-volume work that comprehensively summarizes the knowledge of acupuncture points and is considered the most important Chinese medical manual on acupuncture points. The modern, scientifically reflected form of TCM that is taught at special medical universities in China today is ultimately based on such pre-modern foundations.

Although, according to TCM, acupuncture points are to be thought of as systematically defined points on the surface of the body, every person's body is a bit different. Therefore, traditional Chinese medicine has developed a unique way of measurement. This method uses a fundamentally different form of measurement than the units we use in everyday life, such as centimeters, inches, and so on. The basic unit for measuring the position of acupuncture points is not an absolute length, *but a relative, proportional unit of length, a variable proportional unit of measurement*. This method of measurement, which is therefore also called *tongshen cun* 同身寸 (literal translation: 'own body measurements'), has been developed and standardized over thousands of years.

In the following, some basic relative lengths of this measurement method are briefly described, so that readers can better understand the subsequent analysis and try the measurements on their own bodies. Basically, there are two main units of length in Chinese medicine: (a) thumb measure, that is, the width of the middle knuckle of the thumb of the person being measured is taken as 1 horizontal *cun* (*heng cun* 横寸); (b) middle finger measure, meaning that the distance between the wrinkles of the 2nd and 3rd knuckles of the person being measured when the middle finger is bent is taken as 1 vertical *cun* (*shu cun* 竖寸). The length of the horizontal *cun* is used to determine the location of each acupuncture point in the horizontal direction on the face, chest, and lower back; the vertical *cun* is used to measure each acupuncture point in the vertical direction on the head, chest, and abdomen, as well as the vertical distances on the lower extremities of the upper limbs (Liu and Liu 2009, 16–17). It should be noted that in TCM, these two basic lengths are not used to determine the location of all acupuncture points. They must be used in conjunction with the detailed position of the bones of each measured person and the proportions of the bones—which is also known as the "bone measurement technique" (*gudu fa* 骨度法) (Sui and Meng 2013, 2–4). This personalized method of proportional measurement, based on the partial length of each person's body as the basic length, undoubtedly allows for maximum individual variability in body size. For example, the width of each rib is counted as about 1.6 *cun*, the distance from the forehead to the frontal hairline as 3 *cun*, and so on. Thirdly, there are also further anatomical landmarks that are used in practice for accurate acupuncture point localization (Liu and Liu 2009, 17).

In order to quickly and accurately locate different acupuncture points on different parts of the body, these three measurement methods are often used in combination. Please note that the data analysis performed in this article relies exclusively on measurements derived from the first-mentioned proportional measurement approach. In particular, I refer to the corresponding official national standard (Chinese National Standard GB/T12346-2006 "Names and Positions of Acupuncture Points," 国家标准代码 GB/T12346-2006《腧穴名称与定位》), which is used in modern applications of TCM (Sui and Meng 2013).

It is worth noting that the vast majority of texts on the subject, both ancient and modern, focus on individual acupuncture points or on the specific location of a point and its functions. But what if we could find a completely new perspective from the point of view of music theory? In my view, the structure of the distribution of human

Table 1 Meridial intervals and their equivalent in musical terminology

Ratio	Frequency ratio	Name of corresponding musical interval[2]
1:1	1:1	Perfect unison
1:2	2:1	Octave
2:3	3:2	Perfect fifth
3:4	4:3	Perfect fourth
3:5	5:3	Major sixth
4:5	5:4	Major third
5:6	6:5	Minor third
5:8	8:5	Minor sixth
5:9	9:5	Minor second
8:9	9:8	Major second

acupuncture points on the surface layer of the human body represents a mathematical relationship of intervals, and the frequency with which some of these intervals occur shows a certain regularity. In the interest of space, I have limited the present perspective with regard to two patterns of relationships. The following Tables 2, 3, 4 and 5 contain representative lists of neighboring acupuncture points. These form pairs of three, and their distances to each other are converted into mathematical ratios in terms of the aforementioned finger measures.

Table 1 shows the possible proportions of the ratios of acupuncture points in triple chains of neighboring acupuncture points on the meridians and the names of musical intervals corresponding to these proportions. Related to this, Fig. 1 shows the harmonic vibration ratios of the string of a musical instrument as obtained by dividing the string according to the first five mathematical ratios 1:1, 1:2, 2:3, 3:4, and 3:5 listed in Table 1. These correspond to the the first five musical intervals shown in Table 1.

It should be noted that the data presented in Tables 2, 3, 4, and 5 as well as, correspondingly, in Figs. 2, 3, 4, and 5 represent only a rather small sample of acupuncture points of different regions of the body, and that there are actually more proportional relationships. In the interest of space, these other examples cannot be analyzed here. But the four examples should be sufficient to demonstrate the intervallic proportional relationships embodied in the spatial locations of the body's acupuncture points. Based on this, an attempt can be made to provide a basis for the following intermediate propositions:

(1) Human acupuncture points are distributed systematically in organized patterns over the various surface areas of the segments of the body. This systematic nature of their distribution is also reflected in the mathematical proportions of the distances between acupuncture points. These can be related to the numerical proportions of musical intervals in acoustics.
(2) The intervallic relationship between acupuncture points is universal, both in terms of the close proximity of acupuncture points to each other and in terms

Table 2 Examples of the proportions of acupuncture points on the arms (see also Fig. 2)

Meridian	Acupuncture points*	Proportional value of interval distances	Musical intervals
Lung Meridian of Hand-Taiyin	LU3-LU5: LU4-LU5	6:5	Minor third
	LU4-LU6: LU5-LU6	1:1	Perfect unison
	LU5-LU6: LU6-LU8	5:6	Minor third
Large Intestine Meridian of Hand-Yangming	LI5-LI6: LI5-LI7	3:5	Major sixth
	LI8-LI10: LI9-LI10	2:1	Octave
	LI9-LI11: LI10-LI11	3:2	Perfect fifth
Triple Burner Meridian of Hand-Shaoyang	TE11-TE13: TE12-TE13	2:1	Octave
	EC†-TE11: TE10-TE11	2:1	Octave
	EC-TE8: EC-TE9	8:5	Minor sixth
	WL‡-TE8: TE8-TE9	4:3	Perfect fourth

* All acupuncture point location data in this chapter are derived from Sui and Meng (2013). † Elbow crease. ‡ Wrist line

Table 3 Examples of the proportions of acupuncture points on the legs (see also Fig. 3)

Meridian	Acupuncture points	Proportional value of interval distances	Musical intervals
Stomach Meridian of Foot-Yangming	ST31-PTE*: ST31-ST32	3:2	Perfect fifth
	ST31-ST33: ST32-ST33	5:4	Major third
	ST33-PTE: ST34-PTE	3:2	Perfect fifth
	ST32-PTE: ST33-PTE	2:1	Octave
	ST35-ST37: ST36-ST37	2:1	Octave
	ST36-ST38: ST36-ST37	5:3	Major sixth
	ST35-ST41: ST38-ST41	2:1	Octave

* Patella top edge

Table 4 Examples of the proportions of acupuncture points on chest and abdomen (see also Fig. 4)

Meridian	Acupuncture points	Proportional value of interval distances	Musical intervals
Stomach Meridian of Foot-Yangming	ST19-ST25: ST22-ST25	2:1	Octave
	ST19-ST25: ST20-ST25	6:5	Minor third
	ST19-ST25: ST21-ST25	3:2	Perfect fifth
	ST26-ST29: ST27-ST29	3:2	Perfect fifth
ST*-CV†: ST-KI‡		4:3	Perfect fourth

* Stomach Meridian of Foot-Yangming. † Conception Vessels. ‡ Kidney Meridian of Foot-Shaoyin

Table 5 Examples of the proportions of the head points (see also Fig. 5)

Meridian	Acupuncture points	Proportional value of interval distances	Musical intervals
Bladder Meridian of Foot-Taiyang	HF*-BL4: BL4-BL5	4:3	Perfect fourth
	BL4-BL6: BL5-BL6	2:1	Octave
	HF-BL7: BL5-BL	5:3	Major sixth
	HO†-BL10: BL11-BL10	5:3	Major sixth
	HO-BL9: HO-BL10	8:5	Minor sixth

* Hairline forehead. † Hairline occiput

of distant acupuncture points being located on different segments of the body. The vast majority form harmonious intervallic relationships with each other. In general, the interval proportions of perfect unison, octave, perfect fifths, and perfect fourths seem to be most common. The proportions of distant acupuncture points are not shown in the above tables due to space limitations. But interested readers can do the calculations on their own. As an approximate rule, one might say that the closer the points are to each other, the more "consonant" the proportions tend to be, that is, the more intervals with smaller integers are found (see Table 1); the farther away from each other, the more "dissonant" intervals one tends to find, that is, intervals with larger integers.

(3) Not only ratios between different points on the same meridian, but also the intervallic relations of acupuncture points between different meridians exhibit highly harmonious ratios. For example, the parallel arrangement of the three abdominal meridians shown as solid straight lines in Fig. 4—representing the

Stomach Meridian of the Foot-Yangming, the Conception Vessels, and the Kidney Meridian of the Foot-Shaoyin (see Table 4)—clearly illustrates this. Apart from this example, such trans-meridial relations are not discussed here. When included, an even more complex geometry of intervallic relationships becomes apparent.

(4) It should also be noted that some of the acupuncture points, such as those on the joints, or the distribution of acupuncture points on the chest and, correspondingly, those on the back, are not discussed in this article, and the acupuncture points on the joints are not discussed in this article. This is because they are located on the tissue covering the ribs, spine, and joints. These acupuncture points are related to the bone measurement technique mentioned above, and the data on which this article is based is not related to it.

The acupuncture points observed and analyzed above represent only a small portion of all the acupuncture points of the human body. The study of the proportions or intervals of all acupuncture points to each other would require a comprehensive research project in terms of a completely new perspective of understanding. Such a project would have to involve an interdisciplinary approach, bringing together specialists in TCM, mathematics, and music theory. Only by precisely locating all the acupuncture points, measuring the distances between them, and making accurate calculations could the data obtained then be analyzed in terms of music theory. However, the evidence presented here suggests that a further, more comprehensive verification of the validity of the arguments could lead to the development of an entirely new, *harmonic biological theory* to describe the life processes of the human body.

As explained above, the task of this paper is not to provide a complete theoretical description, but rather to propose a new way of understanding the implicitly harmonic nature of human vital energy in the context of which new directions and methods of research could be developed from an interdisciplinary perspective. As a researcher in the field of music theory, I would also like to point out here the further possibilities of *musicological* comprehension and implementation in view of this phenomenon:

In Chinese medicine, it is generally assumed that the human body consists of a system of twelve main meridians (mirrored along the medial plane of the human body according to its bilateral symmetry), plus the Ren Meridian and the Du Meridian.[3] Put simply, these meridians are like "riverbeds" along which *qi* energy flows to different parts of the body. Each meridian has a different number of acupuncture points. Theoretically speaking, the acupuncture points distributed along a meridian can be imagined as points on an abstract linear string. In this way, the distribution of acupuncture points on a given meridian line, which is given in the standard theory of meridians in Chinese medicine mentioned above, can be accurately mapped. Since it has been shown in the previous tables that there is always some proportional relationship between the acupuncture points in terms of tone intervals, it is possible to imagine that the acupuncture points subdivide each meridian in analogy to different musical scales.

According to known acoustic principles, when a string vibrates, there is the phenomenon of a standing wave. Standing waves form nodes, and the distances between these nodes represent mathematical proportions. Some of these specific numerical ratios are called harmonic interval ratios in music theory. If the meridians of the human body are viewed as "strings" in the transferred sense, and the energy of the body, or *qi*, moves through the meridians in a kind of bio-current, one can hypothesize that the acupuncture points are distributed along the meridians in harmonic proportions in analogy to the nodes of the standing wave of a vibrating string. That is, when *qi* flows through the meridians, waves of different frequencies are generated and standing waves with acupuncture points as their "nodes." Since the length of the meridians varies, the acupuncture points might also reflect the way these "strings" vibrate in different, yet harmonic basic "frequencies" of life energy moving through the body (Fig. 6).

This analogy has both theoretical and practical implications: In theory, the meridians that run through different parts of the body can be viewed as strings that vibrate and carry energy to all parts of the body, making the human body a "musical instrument" with a great number of meridial "strings." According to the basic calculations of music theory, a single string, divided into different proportions, can produce different tones that, when strung together from top to bottom, form a musical scale, and when these tones are played simultaneously or consecutively, they can also form chords or melodies. In this sense, on our "body instrument," there are multiple "voices singing" at the same time,[4] and although our ears cannot hear this polyphony directly, we can "hear" it in the sense of reconstructing it in theory.

Furthermore, this means that each person's body is unique, because each person has a different ratio of body length to body width. This diversity results in each body having a unique "standard pitch." The melodies played in different standard pitches can therefore be considered as music with different "keys" or "tonalities." It follows that the somatic "music" played by each human body is unique in some way.

Understanding the harmonic nature of the human body is of great importance in music therapy. As part of what is known as sound therapy, music therapists are known to treat physical or psychological ailments by using instruments that emit specific tones and intervals to trigger resonances in the living tissue of the body. Based on acupuncture point harmony, such forms of therapy could be developed further by using special "tuning devices" to trigger the resonances of specific acupuncture points also in the sense of their intervallic relations, or by having the patient sing or listen to harmonies specifically tuned to their body, etc.

The Human Body as the Singing Universe

The connection between music and the universe represents an ancient thesis. The idea that the first moment of the world was accompanied by a certain sound can be found in the myths and legends of various ancient cultures, for example, in the creation myths of India, China, Egypt, or in the *Bible*. By analogy, even today's metaphor of the "Big Bang" can be understood as a variant of this image (e.g., Cramer 2013).

Since the time of ancient Greece, famous philosophers who also excelled as mathematicians, music theorists, and natural scientists believed that the universe and music were necessarily connected by an order that could be understood in numerical and geometric terms. Pythagorean and Platonic thinkers used mathematics and music theory as tools to construct a harmonic cosmology that has profoundly influenced the intellectual world to this day. In parallel, ancient Chinese cosmology assigned tones and numbers to the celestial bodies to form a cyclical and balanced world system theory based on *yin* 陰 and *yang* 陽, the 'five elements' (*wu xing* 五行) as the basic elements, and the '(ten) heavenly tribes and (twelve) earthly branches' (*tiangan dizhi* 天干地支) as the basis for calculation. On this foundation, twelve tones were systematically assigned to the twelve months of the year and to the movements of the heavenly bodies. In this context, musical tones were systematically arranged and regarded as sound phenomena of the highest world order, that is, as natural expressions of a 'self-so-ing principle of the heavenly order' (*ziran tianli* 自然天理). This area involves interdisciplinary knowledge in view of the *Yijing* 易經 (*Book of Changes*), Chinese Confucian ritual and music philosophy, ancient Chinese astronomy, and music theory, to name a few of the aspects that I have also covered in my dissertation (Peng 2019).

In this section, I would like to extend the musical-harmonic arrangement of human acupuncture points proposed above to open up a new dimension of thinking, while of course acknowledging that the following is still in a rather nascent state and needs to be confirmed and refined by more scientific observations and calculations. But I believe that if this extended hypothesis is pursued further, it may help us to understand human life from an entirely new angle—namely from a hitherto unrecognized perspective of cosmic life.

The proposed extension of the hypothesis is related to the harmonic cosmology of Johannes Kepler (1571–1630). The German astronomer combined the geometric cosmology of the Platonic school and the heliocentric theory of Copernicus with the precise observations of the planetary orbits by Tycho Brahe. It is often forgotten that Kepler was also a music theorist and that this field was directly related to his astronomical views. In his book *Harmonices mundi* (1619), Kepler proposed that the ratio between the aphelion and perihelion of the planets of the solar system, or the ratio between the perihelion and aphelion of the planets, often correspond to integer ratios. For example, the ratio between the perihelion of Saturn and the aphelion of Jupiter is 1:2. In music theory, this is mathematically equivalent to the interval of an octave. And the ratio between the aphelion and perihelion of Mars is 2:3—a perfect fifth in the sense of music theory, and so on. In this way, Kepler discovered a correlation between the movements of the planets and musical intervals. In his book, he presents his calculations regarding the intervals corresponding to each planetary motion and the musical melodies corresponding to each planetary motion (Fig. 7).

In Kepler's view, the structure of our solar system resembles the performance of a musical composition, or "a polyphonic planetary chorus," if I may say so. According to Kepler, it is of course inaudible to the human ear and can only be understood mathematically by the rational mind. At the same time, as a Christian, he sees this cosmic music in the relationships of the planetary movements as evidence of God's creation, the principles of which can thus be studied in the sense of a *musical order of the universe.*

Kepler's theory is amazingly comprehensive and beautiful. If we put the religious ideas in brackets, we could also say that it represents a very enlightening systems-theoretical approach, even from today's perspective: If one understands the world as a wholeness in diversity, then everything individual in this unity and totality of the world must be coordinated or attuned to it, that is, it must be in universally harmonic relations with everything else. It makes much more sense and is more logical that everything that exhibits some kind of (living) order (instead of arbitrary chaos) is interconnected, rather than everything being random, jumbled, and disconnected. In this sense, everything that follows some kind of order can be understood and grasped as a subsystem from the perspective of a larger systematic whole. Then it is also logical that opposing factors, analogies, and causal connections belong together. Kepler's work shows us that the motions of the planets in our solar system have a harmonious geometric structure. And since we are organisms in the solar system, shouldn't we, as organically embedded beings, and even if we are only a "microscopic" segment of the whole, somehow mirror the harmonious constitution of the solar system as a whole?

A contemporary researcher of Kepler's theory, the German scientist Hartmut Warm, caught my attention with his research. In his book *Die Signatur der Sphären: Von der Ordnung im Sonnensystem* (*The Signature of the Spheres: On the Order of the Solar System*), Warm applies data from modern astronomical measurements to Kepler's theory (Warm 2011). Warm uses a computer program that he wrote for the purpose of analyzing the orbits and movements of the planets in more detail. The orbital data that he refers to include the major semimajor axis, minor semimajor axis, orbital period, orbital velocity, perihelion, and aphelion of each planet. A summary of all the data is provided in Warm (2011, 392–400.) The data are then used to calculate the motions of the planets over very long periods of time in order to reveal the musical proportions hidden in the data and to also represent them in a geometric way. Also from this present-day scientific perspective, the orbits of the planets in the solar system are shown to represent basic geometric shapes, thus revealing the hidden harmonic structure of the celestial orbits. This, in turn, can be translated into musical intervals.

What struck me about Warms's findings and this whole way of thinking was that some of the central celestial intervals correspond to interval ratios of important acupuncture points of the human body. The next table further below shows the ratios of the minor semi-axes of pairs of planets of the solar system (including Pluto) as calculated by Hartmut Warm. I have converted the frequency ratios from his book into musical intervals and named accordingly. It should be noted that these intervallic ratios are not absolutely identical to the frequency ratios but approximate. Hartlieb has rightly pointed out that a dynamic system like the human body, regulated by natural constants, is not a technical regulator. Biological regulators work according to statistical principles, that is, they are accurate only within a certain tolerance range and only as long as the overall structure remains stable (Hartlieb 1969, 215). In view of this, I believe that, given the macroscopic dimension of the solar system and the universe, as well as the microcosmic principle of the individual disposition of human bodies (212), the aforementioned very close approximation includes an acceptable margin of possible variation. In this context, it should be noted that, if one wants to be very precise about the definition of harmonic ratios in music, these have undergone many different calculations that yielded slightly varying results throughout their conceptual history. Against this background, these concepts do not represent an absolutely fixed quantity. The complex developmental history of the perception of musical intervals in different cultural contexts teaches us that, also from a historical and transcultural perspective, these intervallic ratios themselves can be understood in terms of a certain limited range of microvariation.

It is instructive that Warm also presents some of the above relationships in geometric form (2011, 26). Based on his study, I have attempted to compare this data with the harmonic constellations of acupuncture points. In particular, I chose the main meridian through which the body's *qi* energy flows, the Ren Meridian, to compare with Warm's planetary data set. It is important to note that this comparison is not only mathematical, but also involves an enhanced understanding of the meaning and function of acupuncture points and meridians. However, due to space limitations, the relevant context can only be briefly presented as additional background knowledge in this article: In Chinese medicine, there is a fundamental meridian in the center of the human body called the Ren Meridian. It is also known as the 'Conception Vessel,' because it begins in the human reproductive organs and can be described as the source of human life. The course of the Ren Meridian is represented by the midline of the chest and abdomen, and it passes through a number of major organs. According to Chinese medicine, the Ren Meridian serves as the general source for all *yin* meridians in the body, and thus represents a more fundamental level of energy function even than the system of the twelve main meridians of acupuncture.

In my diagram, in Fig. 8, from top to bottom, one can find the representation of a superposition of certain relationships in the movements of the planets Mercury, Venus, Earth, Mars, Jupiter, Saturn, Uranus, Neptune, and Pluto, which are shown in geometric form, with the position of certain acupuncture points along the Ren Meridian: The relationships of the minor semi-axes of the elliptical orbits of these planets to each other, the calculations of which are presented in Table 6, are shown here as circles. For my circular representation of some of these relationships in Fig. 8, I follow Warm (2011, 26), who presents a more complex and comprehensive arrangement. When this information is superimposed on the Ren Meridian, it can be seen that (1) the Ren Meridian passes through the diameters of all these circles and represents their shared axis. (2) In the geometric representation of the abstract positions of all these relationships of the movements of the individual planets to each other, there is a significant, not to say very precise overlap with corresponding acupuncture points (which, of course, also stand in relation to each other). Astonishingly, according to the theory of TCM, these very same acupuncture points along the Ren Meridian represent the most important storage centers for human life energy in the body: CV4,

5, 8, 11, 13, 17, 19, 20 and 23. They are also the most commonly used and important points in the practice of TCM. (3) The minor axis ratios of the planets are very harmonious from the point of view of music theory, with perfect intervals (perfect octave, perfect fifth, and perfect fourth) occurring eight times. (4) These harmonic planetary ratios, when expressed geometrically, overlap with the Ren Meridian and reveal the hidden harmonic intervals of the acupuncture points. (5) In addition to the ratios of the minor semi-major axes, the ratios of the major semi-major axes and the orbital periods of the planets also contain a large number of harmonic intervals, and whether these ratios are analogous to the arrangement of acupuncture points requires further study.

It should be noted that the present exploration is not intended to conclusively prove a causal link between planetary movements and specific human acupuncture points. Rather, it is intended to suggest a new dimension of thinking and new research questions: Is there an implicit harmonic order in the solar system that unites the macrocosm and all beings as its microcosms? If the proportions of the planetary movements and the proportions of the meridian points of the human body can be traced back to the harmonic proportions of music, does the energy cycle of the human body have anything in common with the planetary movements? Do other acupuncture

Table 6 Calculation of the ratios of the planetary minor semi-axes and their approximate musical intervals with the exact calculations according to Warm (2011, 23)

From planet to planet	Music intervals		Small semiaxes of two planets to each other
	Frequency ratio (approximate)	Interval (depending on the harmony level)	
Uranus: Pluto	2:1	Perfect octave	$1.9947 \approx 2$
Saturn: Uranus	2:1	Perfect octave	$2.0121 \approx 2$
Mercury: Venus	2:1	Perfect octave	$1.9094 \approx 2$
Mercury: Mars	4:1	2 perfect octaves	$4.0045 \approx 4$
Saturn: Pluto	4:1	2 perfect octaves	$4.0135 \approx 4$
Earth: Mars	3:2	Perfect fifth	$1.5172 \approx 1.5$
Uranus: Neptun	3:2	Perfect fifth	$1.5683 \approx 1.5$
Venus: Earth	4:3	Perfect fourth	$1.3823 \approx 1.3$
Neptun: Pluto	5:4	Major third	$1.2719 \approx 1.25$
Jupiter: Saturn	9:5	Minor seventh	$1.8335 \approx 1.8$
Mercury: Earth	8:3	Complementary to minor sixth	$2.6394 \approx 2.7$
Mars: Jupiter	7:2	Complementary to Huygens' tritonus	$3.4256 \approx 3.5$
Mercury: Saturn	25:1	Complementary to Chromatic semitone	$25.1519 \approx 25$
Mars: Pluto	25:1	Complementary to Chromatic semitone	$25.2079 \approx 25$

points and meridians have harmonic proportions that are comparable to other data on planetary movements? How should we go about finding and understanding these relationships? Furthermore, we should not limit ourselves to Kepler's or Warm's calculations of the periodic movements of the planets to understand the harmonic similarities between the human body and the celestial bodies. Is it possible to combine this approach with perspectives from quantum physics, string theory, or fundamental cosmic constants to find more similarities between the human body and cosmic phenomena?

Music therapy is generally the use of all forms of music-related activities, such as listening to music, singing, and playing instruments, as a means of treating physical and psychological disorders. Some forms of music therapy today use the principle of sound resonance through passive listening or active engagement with specific sound frequencies. Much research and practice has been done in this area in relation to TCM (e.g., Liu and Yuan 2005; Yu et al. 2009; Jia et al. 2017; Zhang et al. 2018; Weber et al. 2021; Syros et al. 2022). If the body displays a certain harmonic foundation at the level of the meridian points, then it can be seen as corresponding to a musical instrument with harmonic intervals, and therapists may think about methods to make the acupuncture points "resonate" by playing the specific tones and intervals which correspond to the specific points, or by having the person being treated sing or play these tones and intervals by themselves. In a sense, this is comparable to using a "tuning fork" with the right tone and to "tune" acupuncture points, namely to correct inaccurate "tones" and, thus, to achieve a therapeutic effect.

If the analogy between the celestial bodies and the human body is recognized, then the view of a connection between man and the universe ceases to be metaphysical, for this connection is directly represented in our individual physiology; and then it is to be assumed that each human body is in constant resonance with the celestial bodies of the solar system, because our bodies and these celestial bodies basically "play the same music," so to speak. Through this resonance, as living beings, we are in harmony with other beings. And understanding this interconnectedness means a necessary process for human beings to recognize themselves as manifestations of cosmic life—and thus as cosmic beings.

Figures

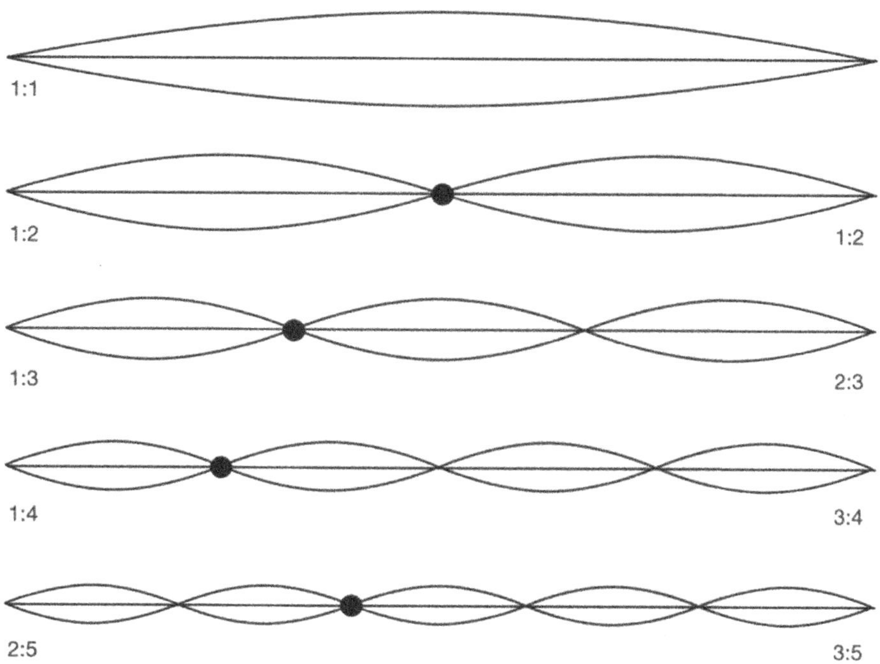

Fig. 1 Harmonic vibrations of strings

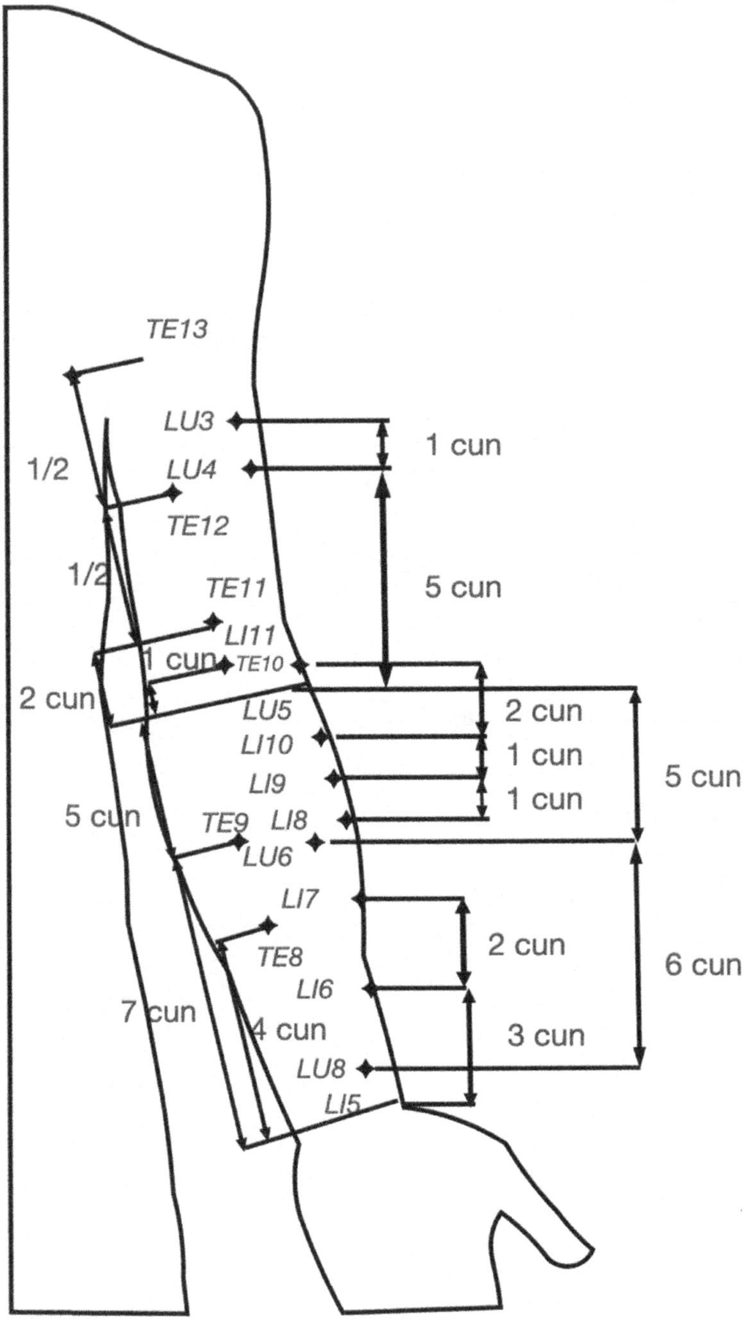

Fig. 2 Examples of the proportions of acupuncture points on the arms (see also Table 2)

Fig. 3 Examples of the proportions of acupuncture points on the legs (see also Table 3)

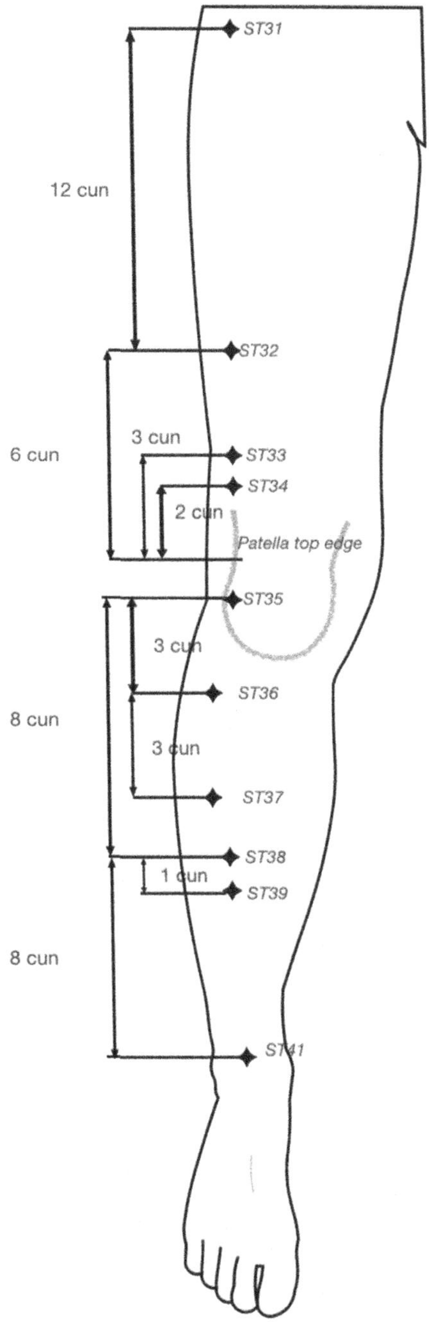

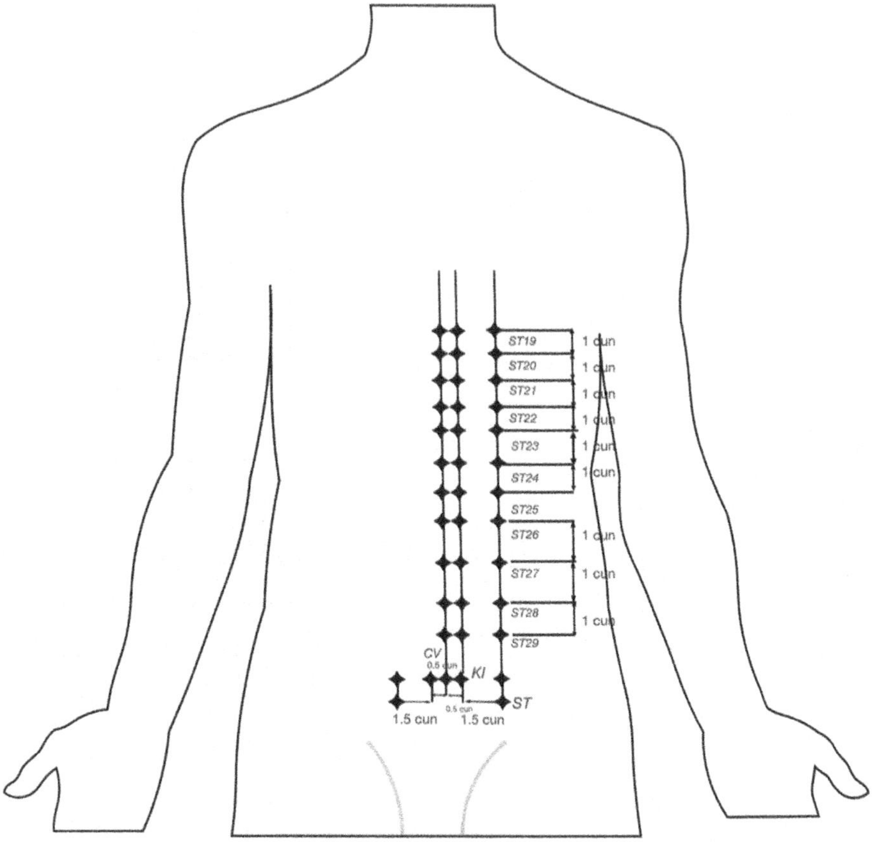

Fig. 4 Examples of the proportions of acupuncture points on chest and abdomen (see also Table 4)

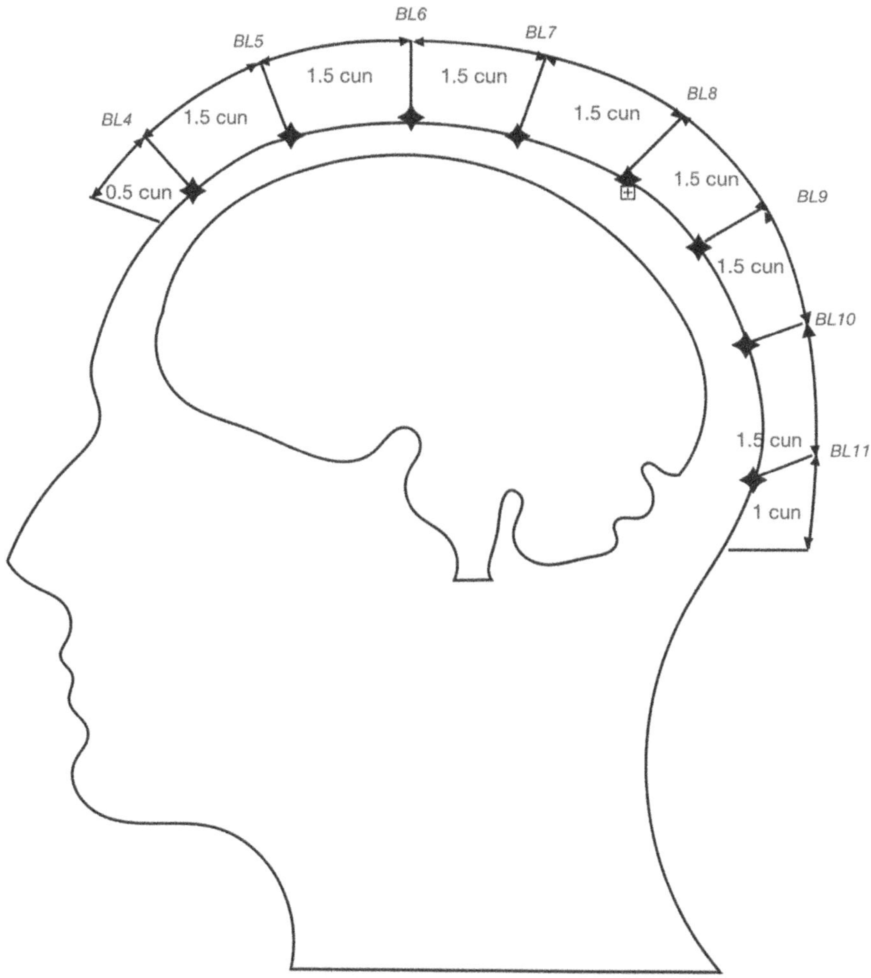

Fig. 5 Examples of the proportions of the head points (see also Table 5)

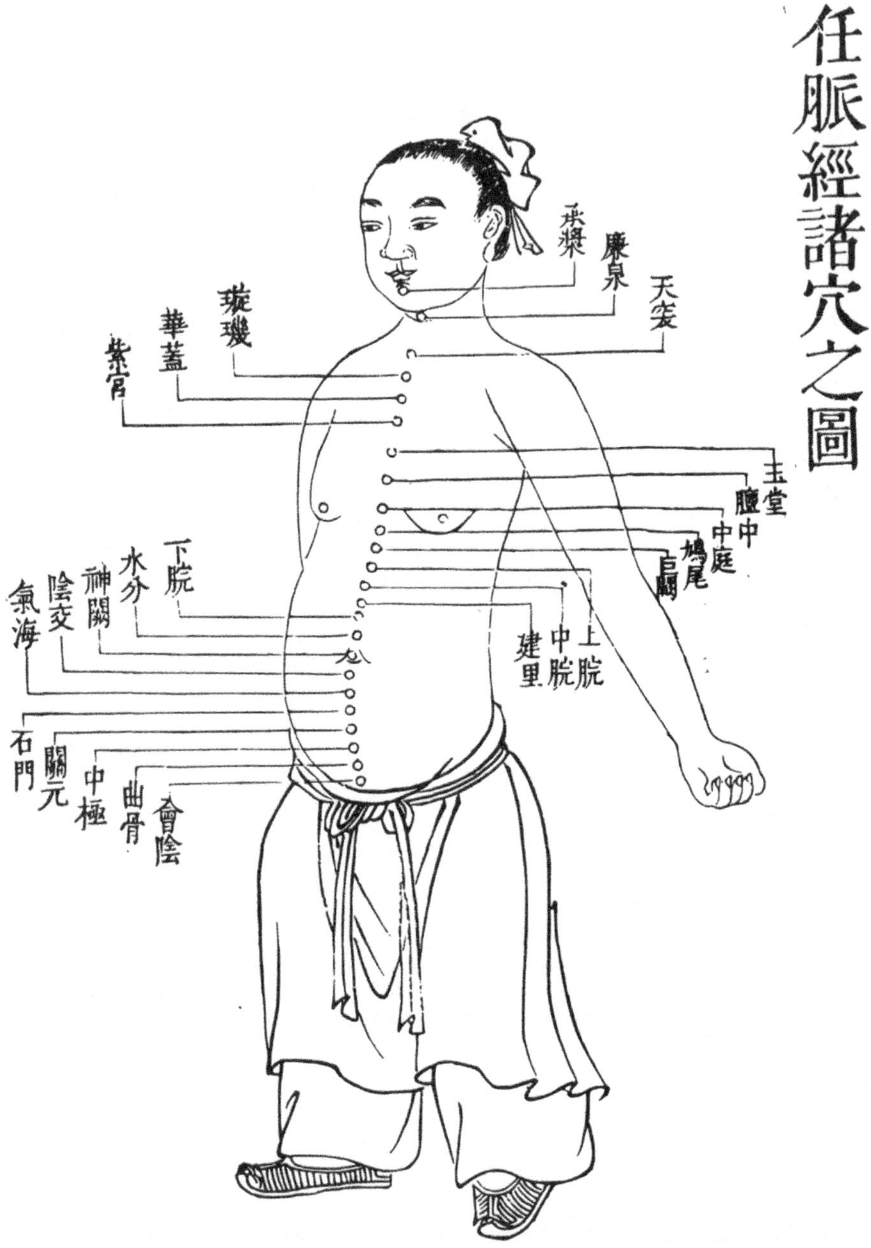

Fig. 6 Can acupuncture points be analogous to nodes of standing waves? "Diagram of the Acupuncture Points of the Ren Meridian" ("任脈經諸穴之圖") (from: *Gujin Tushu Jicheng: Bowu Huibian; Yishudian,* er 72 juan《古今圖書集成·博物彙編·藝術典·第七十二卷》[*Ancient and Modern Illustrated Books: Encyclopedic Collection,* vol. 72: *Skilled Occupation or Profession*; published between 1700 and 1725, completed in 1726]); *Source* Wikimedia Commons, CC Public Domain Mark 1.0. https://commons.wikimedia.org/wiki/File:Imperial_Encyclopaedia_-_Skilled_Occupation_or_Profession_-_pic0285_-_任脈經諸穴之圖.svg. Accessed on February 8, 2023

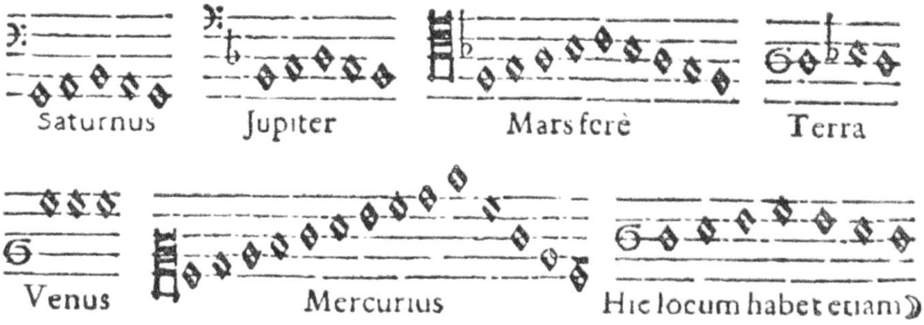

Fig. 7 The "music of the spheres" according to Kepler's *Harmonices mundi* (1619) (quoted in A. Berry, *A Short History of Astronomy* [John Murray, London 1898], p. 190, Fig. 62). *Source* Wikimedia Commons, CC Public Domain Mark 1.0. https://commons.wikimedia.org/wiki/File:Short_history_of_astronomy-Fig_62.png. Accessed on February 8, 2023. *Original source* https://archive.org/details/shorthistoryofas00berruoft

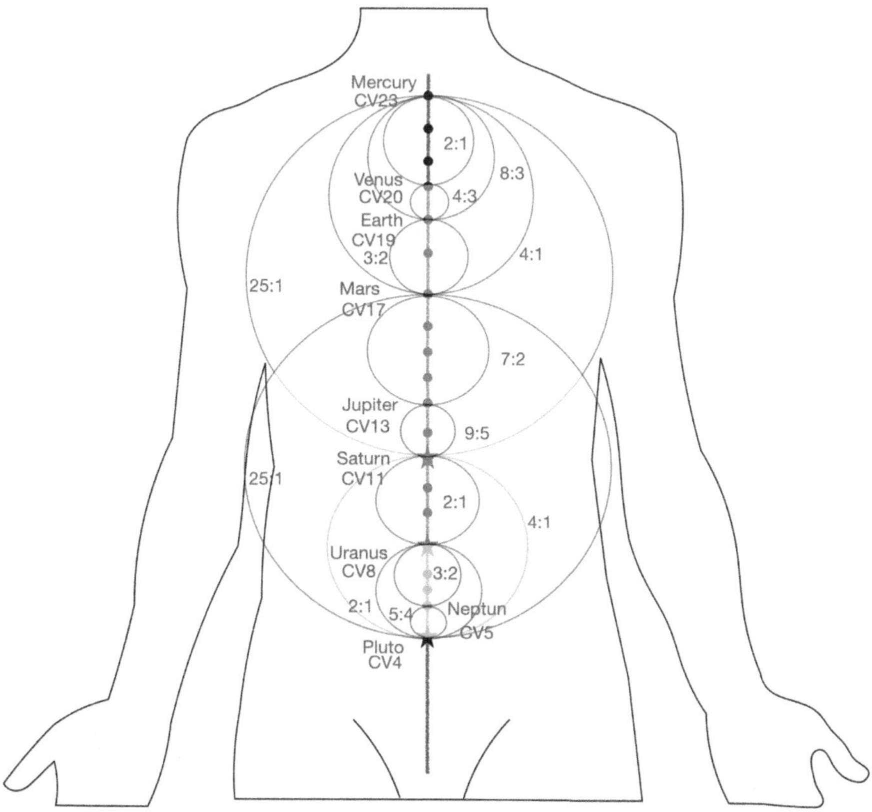

Fig. 8 The ratios of the small semi-axes of the planets are placed on the Ren Meridian and show the correspondence between the planets and the acupuncture points

Notes

1. Not all of these are used in practice, and different schools in different places and at different times have given different numbers. This is also to say that the number of 720 is not to be taken as an absolute standard here. For example, Lu and Needham (1980, 14–15) state that the ancient classic of medicine, *Huangdi Neijing* 黃帝内經, says "that there are 365 acu-points in all, by symbolic correlation with the number of degrees (*tu*) in the celestial circles, to which indeed it likens them, as also to the number of days in the year; and in Han times this number was also supposed to comprehend the number of bones in the human body." The number 720 obviously doubles the number of 360 degrees in a circle. It might be thought of as more or less symbolic. Lu and Needham further state that the *Huangdi Neijing* mentions only 160 acupuncture points by name, while 295 can be deduced from the text in toto. "Until the last half-dozen years the total number was stabilised at 670, but in the burst of further advance which has accompanied the development of acupuncture analgesia, as well as acupuncture treatment of deaf-mutism and paralysis, a considerable number of effective new acu-points has been added, especially in the region of the external ear" (1980, 15).
2. According to legend, Pythagoras discovered the crucial starting point of music theory through listening practice and empirical research: The story goes that Pythagoras passed by a blacksmith's shop, heard the harmonic sounds of hammers, and discovered that the difference in pitch came from the size of the hammers, with larger hammers producing lower notes and smaller hammers producing higher notes. Pythagoras used the simplest numerical ratios of 2:1, 3:2, and 4:3 to explain the intervals of octaves, fifths, and fourths. Harmony was thus rationalized in the form of numerical ratios. This led to the gradual development of a system of music theory, according to such legendary accounts.
3. The number of 14 meridians represents the most common system. There are also systems that include further meridians or acupuncture points.
4. In classical European art music, the most complex piece for a solo instrument has no more than eight voices. So, one can understand how "gloriously majestic" it is that every human body can "play" such a complex "music" of life—an inaudible, cosmically embedded "music" that is itself the somatic precondition of every human artistic expression!

References

A.C. Ahn, J. Wu, G.J. Badger, R. Hammerschlag, H.M. Langevin, Electrical impedance along connective tissue planes associated with acupuncture meridians. BMC Complement. Altern. Med. **5**, article 10 (2005). https://doi.org/10.1186/1472-6882-5-10

Aristoteles, *Metaphysik; Erster Halbband: Bücher I(A)–VI(E)*. Griechisch-Deutsch, Neubearbeitung der Übersetzung von Hermann Bonitz, mit Einleitung und Kommentar herausgegeben von Horst Seidl, griechischer Text in der Edition von Wilhelm Christ. Philosophische Bibliothek, Band 307 (Felix Meiner, Hamburg, 1989)

J.G. Cramer, The sound of the big bang: Planck version (2013), http://faculty.washington.edu/jcramer/BBSound_2013.html. Accessed 4 Feb 2023

K. Hartlieb, Naturkonstanten in Struktur und Funktion der menschlichen Stimme und Sprache. Folia Phoniat. **21**(3), 211–238 (1969). https://doi.org/10.1159/000263252

C.-H. Hempen, *Dtv-Atlas Akupunktur*, 9th edn. (Deutscher Taschenbuch Verlag, München, 2009)

Huainanzi,《淮南子》, Chinese Text Project (n.d.), https://ctext.org/huainanzi/ens. Accessed 3 Feb 2023

C. Jia, H. Zhang, G. Ni, Y. Zhang, B. Su, X. Xu, 贾澄杰、张宏如、倪光夏、张一楠、苏彬、徐新蕾, 头针加音乐联合康复治疗脑卒中后痉挛性偏瘫: 随机对照研究 [Spasmodic hemiplegia after stroke treated with scalp acupuncture, music therapy and rehabilitation: a randomized controlled trial]. 中国针灸 **37**(12), 1271–1275 (2017). https://doi.org/10.13703/j.0255-2930.2017.12.005

H. Kayser, *Lehrbuch der Harmonik* (Occident, Zürich, 1950)

H.M. Langevin, Acupuncture, connective tissue, and peripheral sensory modulation. Crit. Rev. Eukaryot. Gene Expr. **24**(3), 249–253 (2014). https://doi.org/10.1615/CritRevEukaryotGeneExpr.2014008284

H.M. Langevin, N.A. Bouffard, D.L. Churchill, G.J. Badger, Connective tissue fibroblast response to acupuncture: dose-dependent effect of bidirectional needle rotation. J. Altern. Complement. Medi. **13**(3), 355–360 (2007). https://doi.org/10.1089/acm.2007.6351

H.M. Langevin, J.A. Yandow, Relationship of acupuncture points and meridians to connective tissue planes. Anat. Rec. (New Anat.) **269**, 257–265 (2002). https://doi.org/10.1002/ar.10185

G. Liu, L. Yuan, 刘刚、袁立霞, 针刺配合音乐疗法治疗阿尔茨海默病的临床观察 [Clinical observation on acupuncture combined with music for treatment of Alzheimer disease]. 中国针灸 **25**(6), 390–392 (2005)

S. Liu, S. Liu, Z.-F. Wang, Y.-S. Su, R.S. Ray, X.-H. Jing, Y.-Q. Wang, Q. Ma, Somatotopic organization and intensity dependence in driving distinct NPY-expressing sympathetic pathways by electroacupuncture. Neuron **108**(3), 436–450 (2020). https://doi.org/10.1016/j.neuron.2020.07.015

S. Liu, Z. Wang, Y. Su, L. Qi, W. Yang, M. Fu, X. Jing, Y. Wang, Q. Ma, A neuroanatomical basis for electroacupuncture to drive the vagal-adrenal axis. Nature **598**, 641–645 (2021). https://doi.org/10.1038/s41586-021-04001-4

Z. Liu, L. Liu (eds.), *Essentials of Chinese Medicine*, vol. 2: *Clinical Fundamentals in Chinese Medicine* (Springer, London, 2009). https://doi.org/10.1007/978-1-84882-593-2

G.-D. Lu, J. Needham, *Celestial Lancets: A History and Rationale of Acupuncture and Moxa* (Cambridge University Press, Cambridge, 1980)

B. Peng, *Musik als Harmonie von Himmel und Erde – Zhū Zǎiyù (1536–1611) und seine Musiktheorie.* Dissertationsschrift, Universität Heidelberg (heiDOK, Heidelberg, 2019), http://www.ub.uni-heidelberg.de/archiv/26815. Accessed 3 Feb 2023

H. Sui, H. Meng (eds.), 睢海明、孟宏 (编),《最新国家标准针灸穴位便携图》[*The Latest National Standard Acupuncture Points Portable Chart*] (中国中医药出版社, 北京, 2013)

A. Syros, P. Kotlia, G. Fotakopoulos, Preliminary findings from an acupuncture and experiential/traditional music therapy during the standard care of rehabilitation exercise program for recovery on post-stroke upper limb dysfunction. Int. J. Neurosci. **132**(11), 1110–1117 (2022). https://doi.org/10.1080/00207454.2020.1860972

L. Ulloa, Electroacupuncture activates neurons to switch off inflammation. Nature **598**, 573–574 (2021). https://doi.org/10.1038/d41586-021-02714-0

P. Unschuld (ed.), *Huang Di Nei Jing Su Wen: Nature, Knowledge, Imagery in an Ancient Chinese Medical Text*; with an appendix: *The Doctrine of the Five Periods and Six Qi in the Huang Di Nei Jing Su Wen* (University of California Press, Berkeley, 2003)

H. Warm, *Die Signatur der Sphären: Von der Ordnung im Sonnensystem* (Keplerstern, Hamburg, 2011)

A. Weber, S. Busbridge, R. Governo, Evaluation of the efficacy of musical vibroacupuncture in pain relief: a randomized controlled pilot study. Neuromodul. Technol. Neural Interface **24**(8), 1475–1482 (2021). https://doi.org/10.1111/ner.13281.

R. Wilhelm (ed.), *Das Geheimnis der goldenen Blüte*: *Ein chinesisches Lebensbuch.* Übersetzt und erläutert von Richard Wilhelm, mit einem europäischen Kommentar von Carl Gustav Jung (Dorn, München, 1929)

R. Xu, Y. Zhang, 胥荣东、张永旺, 关于‘穴位’一词英文翻译的看法 [Reflections on the English translation of the term ‘xuewei 穴位’”] 中国针灸 **25**(8), 598 (2005)

H.-B. Yu, Y.-F. Liu, L.-X. Wu, Acupuncture combined with music therapy for treatment of 30 cases of cerebral palsy. J. Tradit. Chin. Med. **29**(4), 243–248 (2009). https://doi.org/10.1016/s0254-6272(09)60074-1

H. Zhang, N. Xu, Z. Li, W. Zhang, W. Yi, 张海华、许能贵、李知行、张雯菁、易玮, 针刺配合五行音乐治疗女性抑郁症的临床观察 [Effects on female depression treated with the combined therapy of acupuncture and the five-element music therapy"]. 中国针灸**38**(12), 1293–1297 (2018). https://doi.org/10.13703/j.0255-2930.2018.12.011

Bei Peng (China) is a musicologist, philosopher, and translator who is at home in both Eastern and Western cultures. After a long stay in Germany, she is now working in China at Beijing Normal University at Zhuhai in the Guangdong–Hong Kong–Macao Greater Bay Area. In her work, Peng uses music theory to develop a universal understanding of human existence on an aesthetic level.

Poly-contextural Cornerstones for a Transcultural Philosophy of Cosmic Life

David Bartosch

Abstract In this chapter, important transcultural and multi-civilizational foundations for a comprehensive philosophy of cosmic life are presented from a systematic and at the same time historical perspective. An "anacrusis" regarding the origin of the philosophical term 'cosmic life' is followed by systematic groundwork in relation to Gotthard Günther's concepts of poly-contexturality and trans-classical science. These are extended and complemented by the views of other thinkers. Against this background, the new term 'panenbiotism' ("all-in-life-doctrine") is introduced. Like 'cosmic life,' this term is then used as the basis for a transcultural comparative discussion of positions from ancient Hellenic and Chinese contexts, from South and West Asia, and from Europe. This discussion is organized in the form of three main sections which correspond to three basic themes concerning 'cosmic life.' Each of these is developed along the trajectories of related subtopics. In this sense, the first main section presents several views from various times and places in the context of which (cosmic) life is seen as an expressed principle of incessant self-(re)creation through metabolism and partial self-consumption. In the second main section, the transcultural comparative angle is applied to multiple perspectives and variations on the theme of cosmic life as an integral feature of a vast or infinite universe. Finally, the third main section presents family-resemblant views of cosmic life as an integral aspect of cosmic 'light.' This last section also links the discussion with some unorthodox and very innovative hypotheses of our time. The main thesis of this chapter is that a future philosophy of cosmic life must draw on a variety of philosophical traditions from different civilizations and, in a spirit of unrestricted openness, translate the ultimate inexhaustibility of all scientific endeavors into a more productive intellectual reality within the expanded scope of a new kind of trans-classical concept of science.

D. Bartosch (✉)
Institute of Advanced Studies in Humanities and Social Sciences, Research Institute for Globalization and Cultural Development Strategies, Beijing Normal University at Zhuhai, Zhuhai 519087, China
e-mail: david.bartosch@bnu.edu.cn

Budapest Centre for Long-Term Sustainability (BC4LS), Budapest, Hungary

D. Bartosch et al. (eds.), *Towards a Philosophy of Cosmic Life*,
https://doi.org/10.1007/978-981-99-2131-7_7

Keywords Cosmic life · Transcultural comparative philosophy · Poly-contexturality · Trans-classical science · Biocentrism · Panenbiotism · Philosophy of life

Introduction

If the eye of the heart is open, in each atom there will be one hundred secrets.

Attar of Nishapur

Overview

The present investigation is a philosophical one, also because, *before* addressing the main topic of 'cosmic life,' the preconditions and possible approaches to addressing this topic are examined on a very basic level. The following subsection "Preliminary Considerations" of the present introductory segment thus starts in view of the history of the expression 'cosmic life.' And it provides an introduction to the problem of how to enable a pluralistic global philosophy of cosmic life, which allows for the integration of a pluri-cultural foundation in many histories of thought and also in view of diverging systematic approaches. The subsequent subsection "Poly-contexturality in the Theoretical Exploration of Cosmic Life" directs the attention to the philosophy of Gotthard Günther, and explores some of his special philosophical terms, like, for instance, 'classical' versus 'trans-classical' science, or the idea of (poly-)contexturality in correlation to 'life' (and 'death'). Günther's foundation is extended in the third subsection of this introductory part. It is titled "Extending Classical Science Toward Transcultural and Trans-classical Views." After making reference to other thinkers as well, "Toward a Poly-contextural Trans-classical Panenbiotism," introduces a *new philosophical term*. 'Panenbiotism' ("all-in-life-doctrine") designates a new perspective which forms the basis of the following investigations and considerations. Its concept is further prepared and explored in the final subsection "'Cosmic Life' from a Transcultural Point of View."

The first main section following the comprehensive introductory segment reconstructs historical perspectives on the understanding of "Cosmic Life as an Unceasing Transformative Self(re)creation" from various times and places. It opens with "Thales of Miletus," followed by a longer subsection "Uddalaka Aruni, with Comparative Remarks on Other Thinkers." In the end, the positions of this ancient Vedic thinker are also discussed in the context of more recent views of philosophers and scientists mainly from German-speaking contexts of the 19th and 20th centuries. The subsection "Daodejing, Liji, and Wang Yangming" explores the same basic understanding in relation to Chinese traditions of philosophy. In various forms and traditions, the panenbiotic worldview becomes apparent as a shared foundation of ancient Greek, Vedic, and Chinese thought. The perspective on Chinese traditions is deepened in the subsection "Yijing (Book of Changes)." After that, the pendulum of the investigation swings back to Western Asia in the subsection "Zoroastrian Views and Heraclitus of Ephesus," followed by an "Interim Summary."

The second main segment of this chapter is dedicated to a transcultural and historical exploration of views regarding "Cosmic Life and the Question of its Infinite Extension and Distribution." The first subsection "Numbers as an Expression of the Infinite in Daoist and Pythagorean Philosophy" reviews the somewhat congruent or even shared foundations of Pythagoras's and Laozi's "numero-*logical*" speculations about the foundations of the world. This is followed by a rather extensive overview on "The Idea of an Infinite Distribution of Life in Hellenic, Roman, and Chinese Thought." In contrast, also the historical foundations of a one-world theory (as an exclusive location for life) in Aristotle's philosophy are discussed. In all of this, Günther's terminology of classical and trans-classical science is applied and developed further. After that, the next subsection "Nicolaus de Cusa on the Idea of an Infinite Distribution of Extraterrestrial Life" unfolds a very detailed path toward the reflections of the Early Modern Age in Europe. It is followed by remarks on "Montaigne's View as an Implicit Forerunner of Poly-contexturality?" and another overview on relevant positions "From Bruno, Galilei, Descartes, von Guericke, and Others to Newton and Kant."

The third main section is titled "Cosmic Life and Cosmic Light." It explores a third perspective on the idea of cosmic life, which, so to speak, shines through in the history of world philosophy at several times and in various regions and philosophical cultures until today. The first subsection, "Historical Foundations: Ancient Egyptian, Chinese, and Celtic Views," functions as a kind of preparatory stage. It is followed by an overview in regard to "(Neo-)Platonic Traditions." After that, the perspective shifts away from the global histories of philosophies to a more systematic angle on the present-day discourse. This is also reflected in the next subsection's title "More Systematic: Transcultural Foundations and Innovative (Unorthodox) Approaches." It is followed by "More Thoughts in View of Niklas Luhmann and Gotthard Günther" and the topic "Photons, Life, and Consciousness." Here, 'life' is defined as an aspect of cosmic 'light' in the wider sense. What follows are two more subsections: "Biophotons: Fritz-Albert Popp" and "Connecting Dots in the Light of Solar Astronomy: Attila Grandpierre." They are both dedicated to paradigm-breaking and, thus, contested—and yet extremely relevant and highly innovative theories with regard to the question of 'cosmic life.'

The "Closing Remarks" can be read as a little manifesto in favor of the trans-classical, poly-contextural understanding of philosophy—and thus for more openness in view of a manifold of divergent perspectives on the shared problem of how to understand life in a universal way.

Preliminary Considerations

Nature is all at once.

Johann Wolfgang von Goethe

The term 'cosmic life' was coined by the geologist and paleontologist Heinrich Georg Bronn (1800–1862), the first German translator of Darwin's *Origin of Species*. Bronn uses the German term *kosmisches Leben* (Bronn 1841) in the contemporary sense of astronomy, thereby also viewing the universe as a (cosmo-historical) *process*. In a derived sense, he describes Earth's geological, atmospheric, and aquatic, etc., systems as processes of *tellurisches Leben* (telluric life) (Bronn 1841). The concept of the telluric (from Latin *tellus* 'Earth's surface') has a long historical backstory in Greco-Roman (Mediterranean), Oriental (West Asian), and Occidental (European) traditions of thought (Hoppe 1998). Based on his conceptual extension of 'life' in the sense of what he means by the cosmic and telluric evolutions from an overarching process philosophical perspective, Bronn defines *organisches Leben* (organic life) as the "third [evolutionary] stage of life" (*dritte Lebensstufe*) (Bronn 1843, v, tr. and insertion in brackets DB) and *intellektuelles Leben* (intellective life) as the "fourth level of life" (*vierte Lebensstufe*) (Bronn 1849, 977). Please note that Bronn uses the word *Leben* (life) in the sense of a universal cosmic evolution in four stages of progressive processes building on each other. These represent the self-unfoldment of a self-organizing complexity increase. The scope of Bronn's concept of 'cosmic life' includes astronomical, telluric, organic, and intellective processes.

Moses Hess (1812–1875) most likely adapts the new expression *kosmisches Leben* (cosmic life) (Hess 1877) from Bronn, but he uses it to develop his own *universal concept* of cosmic life. Hess views the "*organic* life process [on celestial bodies] […] [as] a richer, more complicated movement of the general, *cosmic* life process" (Hess 1877, 125, tr. and insertions in brackets DB). In Hess's view, the cosmic life process manifests itself in the formation of stars and planetary systems. Hess even writes of a "cosmic embryology" (Hess 1877, 115), and he takes the *Ableben* (withering away) of stars and star systems in the extended *literal* sense. He comes to the following conclusion:

> All of [...] [the] processes [in the universe] prove that the cosmic action and reaction centers in their first development are subject to the same laws as the organic nuclei. The condensations and combinations of the dynamic centers of gravity are at the same time sources of heat and life.—All embryological processes owe their existence to the same movements of a centripetal action, which forms the substance of life, and a centrifugal reaction, which develops it.—It is just that the more the forming and developing action and reaction progresses, the richer life becomes with regard to its internal self-differentiation.—After all cosmic elements and states of aggregation have formed, and the binary chemical process has been fully unfolded, the organic life process occurs on the cooled-down surfaces of planetary bodies. These are irradiated by larger world bodies still existing in their full power. [...] Similarly, only after the complete formation of the organic sphere of life, the self-conscious or social [sphere of life] appears. At this point, the latest and highest entities of the series of organisms begin to develop their moral and intellectual

> dispositions, which originate from the organic sphere.—So from the simplest and most universal centers the most combined and individual ones are unfolding. In accordance with the same laws to which the cosmic atoms owe their existence, groups of worlds and series of organisms, world-historical peoples and individuals emerge. In a condensed form, the latter inherit the genius of social mankind, the "world spirit," within themselves. Everywhere, transitions fill the apparent gaps, from one sphere of life to another, as well as from lower to higher levels within [each of these spheres of life]. (Hess 1877, 125–126, tr. and insertions in brackets DB)

The task of the present book is to reexamine the philosophical problem horizon of cosmic life. This has to be done in a new and timely fashion and calls for three things: a proper philosophical foundation, the inclusion of innovative, up-to-date science, and an open mind. To provide such a foundation, proper systematic reflection is necessary. John B. Cobb, Jr.'s contribution to this volume is of utmost importance, because it provides an important guiding light for such systematic reflection. In addition, N. Chandra Wickramasinghe makes the very important point that, despite supporting evidence, highly innovative science can be inhibited by a sort of implicit persistence of outdated paradigms. Aristotle's once influential beliefs in an Earth-centered cosmos (see also "Cosmic Life and the Question of its Infinite Extension and Distribution") and in abiogenesis have long been disproven. However, the idea that organic life must be an exclusive and abiogenic "product" of our planet is still prominent, or it evolved into the idea that such abiogenesis, taken as an exclusive origin of life, only takes place on Earth-like planets in habitable zones of solar systems. It is implicitly nurtured by historical layers of an outdated physicalist world view which dates back to the Early Modern Age, when a few scientists came to view the universe as a sort of "clockwork" and the body as a mere mechanism or "automaton." Unfortunately, on the basis of today's updated vulgar materialist, or rather 'ontological naturalist,' version of this world view, alternative ideas of possible origins of living systems (on Earth) or even more extended hypotheses of what life could mean under the background of cosmic dimensions still often are not considered worthy of discussion.

Amongst other things, this is also to say that, *at least implicitly*, systematic approaches in philosophy and the sciences always also have to be viewed as manifestations of their particular preceding histories in contexts of specific philosophical and scientific traditions (in certain geohistorical contexts). Against this background, we should strive to raise the historical and transcultural awareness for our topic. In view of the many *possible* facets of 'cosmic life,' it will be important to further a very open and inspiring, interdisciplinary discussion atmosphere—one that even allows for "risky bets." Moreover, the search for a philosophical foundation concerning the philosophical exploration of a topic like this has to include related prospections in the *world history* of philosoph*ies* (Scharfstein 1998 et al.). This topic should be viewed from as many different angles as possible. In my view, this also includes the necessity to explore a wide range of non-European and Western traditions of philosophy. Upon closer inspection, one will see that various universal or cosmic conceptions of 'life' have emerged in different civilizations and places.

One has to recall that we can learn from Chunyou Yan's contribution to this book that every "*holistic*" view is still a *finite* macro perspective of *someone* (in a particular context). No one can have a *complete* understanding of such a vast and multilayered topic like 'cosmic life.' What we can do, however, is to include a *diverse* spectrum of many philosophical foundations and macro perspectives of many important thinkers of as many times, cultures, and civilizations as possible. By doing so, the possibilities of looking at our topic can be vastly expanded—to generate a more diversified and pluralistic foundation of historical substructures and directions for our systematic thinking. In my opinion, such a pluralistic foundation will enable us to succeed at reflecting on the massive amount of information that our topic entails. There is no place for any academic "mono-culture" or dogmatic one-sidedness here. And there is no way around what the author will call a *trans-classical* approach here (see further below). A future philosophy of cosmic life will have to be *multi*-disciplinary, multi-methodological, multi-cultural, multi-lingual, etc.

Now, if it is the goal to develop a truly *global philosophy of cosmic life* and to approach the question from the many angles of humanity's cultures, wisdom traditions, and knowledge systems themselves, we need to prepare ourselves accordingly. The topic of cosmic life is a truly "all-encompassing" one. Only in this understanding can it be explored properly. One has have to acknowledge that human reason is a "network of heterogeneous, yet multifariously interwoven, not only mutually exclusive but also overlapping fields of rationality, discourses, ways of life, life-worlds" (Waldenfels 1985, 67, tr. DB). The precious 'context(ure)s' (see also further below) and layers of collective philosophical memories underlying these networks hold the cornerstones for our possible edifice of thought.

A many-faceted, open approach toward the question of cosmic life can also be reflected from the angle of a systems-theoretical understanding, according to which

> existence is brought forth by the distinctions of the observer [and] that there are as many domains of existence as kinds of distinctions the observer performs: objectivity in parentheses entails the *multiversa*, entails the notion that existence is constitutively dependent on the observer, and *that there are as many domains of truths as domains of existence she or he brings forth in her or his distinctions* [*or decisions*]. (Maturana 2003, 57–58, italics and insertions in brackets DB)

Furthermore, Humberto Maturana's (1928–2021) understanding also resonates with an important topic in quantum mechanics, that is, *contextuality*.

> Contextuality says that properties of particles, such as their position or polarization, exist only within the context of a measurement. Instead of thinking of particles' properties as having fixed values, consider them more like words in language, whose meanings can change depending on the context. (McCormick 2022)

More than 550 years earlier than Maturana, Nicolaus de Cusa (also: Nicholas of Cusa, Nicolaus Cusanus, Nikolaus von Kues, 1401–1464) presented an idea which partly resembles Maturana's systems-theoretical 'multiversa' of the reflection. In *De coniecturis* (*On conjectures*, 1442), Cusanus states that every human being unfolds their own *microcosmos* (microcosm) of concepts and conceptualities, and thus knowledge (Nicolaus de Cusa 1972, 143). Each of these self-contained personal processes of understanding might be more or less similar to others, but none of them could ever be fully identical with another—an idea which, in a way, partly foreshadows Gottfried Wilhelm Leibniz's (1646–1714) take on the *principium identitatis indiscernibilium*. In analogy to Maturana's multiverse, each of Cusanus's self-unfolding microcosms of knowledge is to be thought of as "*verisimile*" (truth-like) (Nicolaus de Cusa 1972, 113). In the sense of *scientia* (science), and at best, each individual human microcosm of understanding is consistent within itself, but none of them reaches a level of *absolute* perfection or 'truth' (*veritas*).

Under the conditions of finiteness, each scientific observer approximates the infinite complexity of an infinite *universum* (universe). Even its particular phenomena are infinitely complex. Cusanus views the universe as a whole—and even the substantial 'singularity' (*singularitas*) of each concrete thing—as the (infinite) *imago* (image) of an *infinite*, omnipotent principle of creation. While it cannot be characterized in terms of location or time (Nicolaus de Cusa 2000, 45–46), space and time are basic aspects of its 'manifestation' (*apparitio*) (Bartosch 2015, 28, etc.; see also Kreuzer 2000, 184).

From a historical point of view, this insight is astonishing, because, generally speaking, it is in some measure on a par with the problems of non-locality and contextuality in modern quantum mechanics. For Cusanus, the different processes of the cognitive self-unfoldment of the *humana mens* (human mind) are nurturing the personal microcosms of understanding of each individual. They, too, are manifestations of the infinitely creative principle (Wolter 2004, 80). Each represents a unique "imaging process" of the universe in finite terms (Bartosch 2015, 483–84). Therefore, each of them represents a 'truth-like' approximation of absolute 'truth.'

Maturana's statement and his general understanding bear a certain affinity to Cusanus's position. Of course, they cannot be absolutely congruent. Contentual differences and divergent theoretical contexts and philosophical backgrounds cannot be ignored. However, it cannot be denied either that Maturana's views "resonate" with those of Cusanus to some extent. The resulting "family resemblance"—Ludwig Wittgenstein's (1889–1951) metaphor (Wittgenstein 2009, 36[e], etc.) is to be read in the transferred sense of the present train of thought here—is a systematic one. As far as one can tell, Maturana was not influenced by Cusanus directly. Against this comparative background, we can apply both positions to themselves and to each other: Both self-reflexive aspects of different 'multiversa' (Maturana) or 'microcosms' (Cusanus) are 'truth-like' in that they are *approximating* the same "permeable problem horizon" (Bartosch 2015, 17–19, etc., see also further below) of understanding per se (regarding 'understanding,' see also Bartosch 2021).

Furthermore, in view of the aforementioned aim to generate a historically pluralistic, transcultural foundation for a future philosophy of cosmic life, we can infer from this that, like in the case of Maturana and Cusanus, different philosophies might exhibit *systematic affinities* without any historical connection (including the absence of historical exchange processes across languages and cultures). Such occurrences of systematic resonances of contents can be distinguished from those cases where thoughts and ideas are related in the sense of actual historical developments (histories of philosophy). These are based on the derivation, transformation, or development of concepts and points of views under the influence of predecessors in the context of philosophical *traditions* (see also Scharfstein 1998, 1). Cusanus himself arrives at his aforementioned epistemological understanding by starting from Plato (428/27–348/47 BCE), who holds the view that *everything that can be said about the universe can only be approximate and in the sense of likelihood* (Hadot 2006, 157).

In this contribution, I probe different histories and cultural contexts of philosophy in view of what I understand as the general problem horizon of 'cosmic life' in the sense of an overarching comparative concept on this occasion—both in view of general systematic affinities and in relation to particular concept-historical traditions.

Poly-contexturality in the Theoretical Exploration of Cosmic Life

As mentioned, the proposed shift toward a diversified foundation of different perspectives (in the sense of interdisciplinarity and transculturality) can also be understood in the sense of a turn from *classical to trans-classical science*. Gotthard Günther (1900–1984) refers to this transformation process in philosophy, humanities, and the sciences as follows:

> A great epoch of scientific tradition is about to end. It has lasted almost two-and-a-half millennia and philosophers and scientists begin to call it the classical period of science. [...] It is possible to trace the distinction between the classical and the *trans-classical* back *to deeply hidden metaphysical assumptions about the nature of this Universe*. (Günther 1979, 283, italics DB)

If we want to advance our understanding of what we mean by 'life,' exactly these metaphysical assumptions have to be queried themselves. Gotthard Günther's thinking is thus deeply related to our philosophical problem horizon of cosmic life. Based on his *poly-contextural logic*, he develops a general understanding that holds the keys for an advanced, culturally and systematically pluralistic discussion. For Günther, the varying *observer-dependent* philosophical views (see also further above, the quote by Maturana) or theoretical "spectacles" of various thinkers and scientists of different times, civilizational contexts, and present-day scientific or philosophical disciplines can be described as *contextures*. It is important to note that Günther is using the adjective 'context*ural*'—and not 'context*ual*.' Günther's 'contexture' means something different from the usual term 'context.' In this regard, Günther finds that

> [w]herever we extricate any two data from this world, we will find that they share in a common contexture and that their relations can be described by a two-valued logic. This test will never fail us. But since we pointed out *that every ontological datum of the world must be considered an intersection of an infinite number of contextures*, the fact that—any two data we choose to describe in their common two-valued relations belong to one contexture does not exclude that the very same data may also—apart from the contexturality chosen for our description—*belong separately to additional and different contexturalities*. (Günther 1979, 297, italics DB)

Basically, this implies a shift from classical science, which is defined by its mono-contextural directionality, to a trans-classical *nexus or field* of possible reflections and coordinated, yet divergent points of views. Here, a contradiction that occurs from one "truth-approximating" point of view or angle (of one particular contexture) does not necessarily have to disturb the perspective of another contexture. This idea can be extended: In general, such a poly-contextural approach allows to permanently enhance (the given plurality of) our views and to integrate them in our thus expanding and more self-differentiating 'microcosmic' (Cusanus/Maturana) paths of thinking. A particular focus can be realized from a possibly infinite number of different angles, namely, as long as each chosen perspective is methodologically supported and logically valid. The more prospective, "truth-approaching" perspectives we can handle and integrate in our thinking, the more advanced our approximation of the (infinite) puzzle will become. An important precondition for this is that a most basic non-knowing—and, consequently, an open situation in view of possible alternatives—will always be acknowledged. This aspect of a higher self-reflexivity of 'knowing non-knowing' is the *conditio sine qua non* in the unfoldment of all our possible knowledge itself. This is what the idea of poly-contexturality can teach us.

This shift from classical to trans-classical reflection (sometimes Günther uses the adjectives 'classic' and 'trans-classic') means nothing less than an epochal transformation in the way that our knowledge is (permanently) forming itself: from *mono-contextural* to *poly-contextural* and, as one might add, "poly-*conjectural*" thinking. This, of course, must have fundamental consequences for our understanding of life—especially if the word 'life' is related to respective holistic understandings of words such as 'cosmos,' 'universe,' etc.

> [T]he turn from classic to trans-classic thinking means that the mono-contextural concept of Reality is abandoned and replaced by a poly-contextural theory of Existence *which makes room for the phenomenon of Life within this Universe*. In a poly-contextural Universe we do not have to consider Life as an element totally alien to inanimate matter, because matter in itself already contains the seeds of Life in its dialectical contraposition of Being and Nihility. [...] [I]t will be necessary to consider all living organism [sic] as poly-contextural structures. For the classic tradition there is a complete break between Life and Death. (Günther 1979, 208, italics and brackets DB)

This understanding somehow resonates with various new perspectives which have been based on many decades of scientific work and respective personal accumulations of valid knowledge, as in the works of John B. Cobb, Jr., N. Chandra Wickramasinghe, Chunyou Yan, Attila Grandpierre, and many others. In the abovementioned sense of Cusanus, these individual perspectives represent sophisticated philosophical and scientific 'microcosms' of (self-)reflection in relation to the authors' respective (inter-)disciplinary backgrounds. In view of 'cosmic life,' and in the sense of poly-contexturality, their views can be valued as in a chain of different, yet equally valid observer-perspectives. In the sense of Günther's notion of trans-classical science, the validity and the epistemological value of a theorem emerge from its own (inherent) consistency, its contents, and contexts. It has to be evaluated under the background of its own systematic parameters. From a more general, philosophical angle, this also holds true with regard to ancient or "pre-modern" or "non-Western" philosophies of course—and even in view of what Jean Gebser (1905–1973) called the archaic, magic, and mythical structures of historically preceding forms of consciousness (Gebser 1973).

Extending Classical Science Toward Transcultural and Trans-classical Views

In this chapter, Gotthard Günther's original foundation of poly-contexturality is extended: When Günther states "that the two-valued relation of subject and object takes place in a multiplicity of ontological places which *cannot* be made to coincide with each other" (Günther 1980, 87), he misses *that it is precisely this difference* between the "ontological places" *as such (the difference in-and-for-itself)* which provides the underlying *unity* (of unity and difference) of all of these (different) self-unfolding subject–object microcosms of understanding. The *difference itself* functions as the *unifying difference* "amongst" *those "ontological places,"* and as *that which permanently "draws" all distinctions (enabling a multiplicity of ontological places), the difference is itself without any difference* (Bartosch 2021, 136; 2022, 116). Thus, it *inherits boundless possibility and is the unmoved center in all (cosmic) "decision-based manifestation" and in our related cognitive processes* (also with regard to the unique path of every observer-based world experience in the sense of Maturana). *The undivided act of differentiation represents the "non-local locality" of an ineffable all-creative infinity that is implicit in each individual act or process of unfolding knowledge*!

Recently, the author has introduced the new term 'implicate logic' to refer to this most basic foundation of *all* thinking (in the sense of cognitive acts of differentiation) (Bartosch 2022). Apart from this new categorization and the related meta-reflection, the underlying insight itself is not new. For example, Nicolaus de Cusa refers to it by introducing the philosophical enigma of the "wall of paradise" (*murus paradisi*) (Nicolaus de Cusa 2000, 43, 71, 79). It symbolizes a marker of distinction. Namely by "walking" *on* the "wall"—which signifies the act of differentiation itself—the wall and the walker's consciousness "create" the two parts of 'paradise' and 'non-paradise,' that is, the inside and the outside of the "wall." At the same time, the wall, thus to be seen as the *conditio sine qua non* of the differentiation, also *connects the two sides.* Both sides are connected *through* their difference (the "wall"). *The "wall" itself, however, is in itself undivided.* The difference *itself* is *undivided*; it has no boundary (in) itself (Bartosch 2022, 116–117). At the end of the twentieth century, the great systems theoretician Niklas Luhmann (1927–1998) explored *the very same implicate logical foundation* of thinking that Cusanus had already referred to (Bartosch 2022, 116–117; Luhmann 2001, 245; see also "Numbers as an Expression of the Infinite in Daoist and Pythagorean Philosophy").

If "every ontological datum of the world must be considered an intersection of an infinite number of contextures" (Günther 1979, 297), this means that all the possible contextures or observer-multiversa of reflection in their respective "ontological places" actually do *implicitly* coincide in the form of *every* ontological datum itself. Every data point is an appearance of the underlying infinite *unity* (through all differences) of a possibly infinite number of contextures.

Under the background of this basic *trans-classical* understanding, any theoretical monopolies must be viewed as detrimental to further progress in terms of philosophy, sciences, and social systems today; there is no absolute truth of (infinite) cosmic life that could ever be fully realized apart from an unceasing discourse of (finite) observers (in the aforementioned sense of Maturana). However, as mentioned, *we can observe the many approaches of different (finite) observers and their paths of thinking of the past (in their different contexts) to obtain a more diversified and therefore broader and clearer picture of the problem horizon of our topic.* The present investigation therefore starts with a prospection of some of the earliest *pre*-classical layers in various traditions of ancient philosophy. These layers in the earliest histories of world philosophy precede the emergence of classical (mono-contextural) science in ancient Greece. Starting from there, the present contribution will then also provide examples for later philosophical positions which are of equal importance for our topic. These, as well, have to be integrated in the unfoldment of a poly-contextural tessellation of the foundations of a trans-classical philosophy of cosmic life in mankind's history of consciousness (see also Gebser 1973). The major task of the following is thus to probe a great variety of layers in the world history of philosophy for a plurality of inherently coherent and comparable views as well as conceptual roots which can function as historical foundation levels (German: *Fundierungsebenen*) of a global discussion on the topic of cosmic life. (One might add that comparability does not mean the absence of differences, of course.) The discovery and integration of complementary elements in the world's traditions of thought is a basic necessity *in view of a trans-classical and transcultural philosophy of cosmic life.*

Classical science (Günther) only accepts the parameters of *one conjecture.* It is mono-theoretical in the sense that, at least for certain period in time, only one theory becomes the paradigm (or should we say "dogma"?) of the respective "scientific belief" (Bartosch 2019, 47–50) in a particular discipline or sub-discipline. Those who follow alternative possibilities of ("heterodox") explanation often risk having their reputations destroyed, being marginalized, or even worse (think of Giordano Bruno). While 'classical science' can mainly be traced to certain Greek foundations, for example, to Plato's and Aristotle's philosophies, the approach of 'trans-classical' science can*not* mean to exclude Greek and later European philosophical elements and foundations, on the other hand. Useful outcomes and methods of prior mono-contextural sciences and historical discourses are to be integrated into innovative poly-contextural chains of trans-classical reflection. If one were to exclude the former aspects of mono-contextural thinking, this would imply an implicit move in the direction of such a mode of thinking itself.

Furthermore, and while taking into consideration other non-European traditions of philosophy, we also have to recall that some of their representatives have deeply influenced *later* developments of European philosophies and sciences. 'Classical science' in Western contexts has greatly benefitted from the (non-classical!) integration of elements of historical non-Western philosophies and sciences. Although these "foreign" elements and their historical function in Western thought have often not been recognized later on (e.g., the sciences and the philosophies of the Islamic Golden Age), one might still say that, at least *implicitly* and on a rather *"subcutaneous" level*, the shift toward a trans-classical and transcultural approach has long been prepared in the history of classical science and Western philosophy itself. It is about time to draw the proper (poly-contextural) conclusions from such an interlinked history of classical science and those (marginalized) influences and undercurrents. A future philosophy of cosmic life also has to be based on all the useful and valid foundations of East Asian, South Asian, West Asian, European, and many more traditions of thinking in other regions of the world. They all provide the intellectual background for a future philosophy of cosmic life.

At this point, the reader might ask: Why should we go back to ancient or at least older philosophical sources (from a transcultural perspective) in this context at all? What's the gain in terms of scientific progress?

We have to be aware that, to a certain extent, the relationship of philosophy to its (world) history (or: its histories) is different from the (progressive) history of the natural sciences. Certain philosophical ideas and most basic problems remain throughout history. Their foundations often present themselves most accessible to our understanding in early texts (of various civilizational contexts). That is why certain elements of ancient thought might still provide valuable starting points and basic ideas for current philosophical discussions. Furthermore, some philosophies and their terminologies have provided historical starting points of specific disciplines in science, humanities, and sociology. As a historical background or origin of these subjects, historical traditions of philosophy have to be seen in their own light and to be dealt with in their own right and in view of their exclusive "dimension(s)" and way(s) of thinking. The word 'philosophy' pertains to a methodological way of thinking in the most basic and most unconditional sense. As such, philosophy does not start from any particular theory (with a particular restricted focus or area of interest) to be verified or falsified (like in the case of science), but, more fundamentally, from the unbound principle of *knowing non-knowing* itself (Bartosch 2021, 2022).

Carl Friedrich von Weizsäcker (1912–2007)—both a world-class scientist and a renowned philosopher—states that when he reflected on the question of the basic interpretation of complex mathematical theories in his scientific field (quantum physics), certain very basic meanings could often only be comprehended by going back to very original starting points in the early history of *philosophy*: "So my experience was that I was unable to understand quantum theory, and to this day I am unable to understand quantum theory, unless I understand Plato" (von Weizsäcker 1971, 378, tr. DB). Von Weizsäcker's remark is in certain resonance with Alfred North Whitehead's (1861–1947) statement that the history of Western philosophy is nothing more than footnotes to Plato (Whitehead 1978, 39). Both von Weizsäcker and Whitehead did not refer to Plato in the sense of a theory, but to his "form" or way of thinking, which is based on Socrates's (c.470–399 BCE) insight: "I neither know nor think that I know" (Plato 1892, 113–114). This foundation of philosophy does not have to do anything in particular with the (idealist) worldview that Plato derived from it. It is also not a particular feature of Greek philosophy. The Chinese master Confucius (Kongzi 孔子, 551–479 BCE) had discovered it at an earlier date (Bartosch 2022, 108).

With regard to Western philosophy, it is fair to extend Whitehead's remark to other philosophers and even non-idealist positions which have come forth from the same basic insight of knowing non-knowing. In one of his last seminars on natural philosophy, which was held in 1981, Gotthard Günther extended this idea in the sense that the history of Western philosophy and science is to be understood as a struggle between interpretations of Plato (incl. Aristotelian systematics) and materialist philosophies (including mechanistic-reductionist tendencies). On a poly-contextural basis, we can learn to see these divergent philosophical world views as alternatives which do not have to—and which actually cannot be decided. They are just different views of "seeing" or interpreting manifestations of a non-conceptualizable (infinite) creativity at the foundation of our experience and existence. And if we leave the Eurocentric specs aside, we can say that philosophy as such, that is to say, 'philosophy' as an expression of knowing non-knowing, is, of course, not just footnotes to Plato, but a living and fertile "forest" of wisdom which has grown out of the seeds that many great minds have produced in almost all corners of the inhabited world. To provide a fitting example to complement von Weizsäcker's "Platonism," we can think of the Danish physicist Niels Bohr (1885–1962): When he was knighted in 1947, he had to design a family crest for himself. He chose the Chinese *taiji* 太極-symbol. To Bohr, this ancient symbol meant the best representation of his understanding of the wave–particle duality in quantum physics (Aaserud, n.d., 71–72).

Toward a Poly-contextural Trans-classical Panenbiotism

Furthermore, we also have to differentiate in view of the general terms 'idealism,' and 'materialism' here. For example, in view of Western materialist thinking, we can continue to develop Günther's own implicit extension of Whitehead's famous "footnotes-to-Plato-statement" by referring to Ernst Bloch's (1885–1977) insights into the history of materialist philosophies: Not all historical and more recent materialist world conceptions can also be categorized as physicalist or mechanistic, that is, not all forms of materialism marginalize or even deny the importance of a spiritual or non-material aspect—or the "life in matter," as one might put it. The philosopher Bloch, who defined himself as a Marxist materialist, was also influenced by the same contemporary quantum physics that led Weizsäcker and Bohr in the direction of various philosophical traditions of the East and the West (for interpretative purposes). Against that background, also Ernst Bloch vehemently *opposes* those *physicalist or mechanistic* materialist views which merely advocate "*Klötzchen-Materie*" (building block matter) (Bloch 1972, 309) as the sole "ingredient" of the universe.

In Bloch's view, which also drew inspiration from thinkers like Giordano Bruno (1548–1600) and many others, the material universe is inherently *self-motivated*. By necessity, it includes the principle of an all-encompassing *logos*, and it is an expression of *a metaphysical principle of hope* (Bloch's *Prinzip Hoffnung*) which is inherent in (this non-physicalist, living cosmic) matter (which is always in motion and evolving). Bloch's principle of hope is not material itself. Bloch develops his view in a historically supported manner: He takes his first step with the originator of process thinking in ancient Greece, Heraclitus of Ephesus (c.520–c.460 BCE), and then unfolds his dialectical reflection on a universally living *Logikon in der Materie* (logicon in matter) (Bloch 1972, 47) alongside many stations in the history of philosophy in Europe, the Mediterranean region, and West Asia. His philosophical investigation attains a status and sophistication which is way more advanced than other—quasi "gnostic"—attempts which "localize" an allegedly *separate* idea-principle, so to speak, "behind," "above," or "apart from" a seemingly treacherous or even "evil-ridden" world experience—as *Vulgärplatonismus* (vulgar forms of Platonism), that is, the direction of crude yet common misinterpreters of Plato, has tried. The point is that the infinite *cannot be separate(d) from anything*, in particular not from *any* of its individual manifestations (in the sense of processes of its all-creative self-unfoldment).

From an extended perspective on poly-contextural logic in the earlier-mentioned sense, the either-or decision between *valid* interpretations of Plato or, to refer to a modern idealist philosopher, Hegel (who saw the Idea as *a self-enfolding, self-evolving act*, that is, as a reality-generating process of self-translation), and—on the other side—the most sophisticated *organicist or naturalist materialist* ideas, like, for instance, those of David of Dinant (c.1160–c.1215) or, in the 20th century, Ernst Bloch, *is no longer necessary*. Philosophically speaking, one is presented with an aporetic situation here: Can the cosmic becoming of an infinite space be understood as the appearance of an infinite, living mind? Isn't such an intellectual vision (of infinite cosmic mental unity) to be considered as a subjective reflection under the conditions of finite consciousness? Would not this subjectivity (finiteness as a precondition for reflection), in turn, presuppose the reality of an infinitely distributed matter-energy (as a foundation for any physical "localizing" and "timing")? Does it make sense to separate an infinite mind (and its permanent becoming) from an infinitely self-transforming process of matter-energy? I think not. Would both not have to be taken as, so to speak, (two "sides" of) the same "infinite coin"? Yes—namely, in the sense that we cannot decide in favor of one or the other, as I said earlier. We can think of Cusanus's "transgressing" metaphor of an infinite mirror: On the one hand, it symbolizes a universal intellective principle (Nicolaus de Cusa 1959, 48–49), νόησις νοήσεως (*noesis noeseos* 'thinking of thinking'), while, on the other hand, the very same metaphor also implies a polished *material surface*.

It seems that, at least in their most sophisticated developments, both basic views ('idealism' and 'materialism' in the sense of basic directions with many facets and variations) finally tend to coincide: They represent the *same* infinite *panenbiotic* universe—a universe which is so infinitely complex that its underlying ontological principle of poly-contexturality will never grant us the opportunity to reach an end of understanding. The two basic standpoints—either 'mind over matter' or 'matter over mind'—represent two opposing perspectives on the same absolute foundation of Being. This can be referred to by way of the author's new word creation *panenbiotism* (pan-en-biotism). It is composed of the Greek terms πᾶν (all) and ἐν βίῳ (in life). This new term is inspired by the expressions (1) 'panentheism' (from πᾶν [all] and ἐν θεῷ [in God]) and (2) 'biotic community.' The term *Panentheismus* (panentheism) represents a rendering of Karl Christian Friedrich Krause's (1781–1832) German term *Allingottlehre*—a special concept by an innovative (and, unfortunately, much underrated) thinker, who more or less implicitly influenced the discipline of process philosophy during the 20th and 21st centuries (Walker and Wickramasinghe 2015, 39). For the second term 'biotic community,' one can refer to the overview by Landen (2003). The semantics of the term 'panenbiotism' is alluding to a cosmic, non-(Earth-)centered perspective on life, in which (conscious and self-organizing) *Life* does not mean a marginal phenomenon. According to the panenbiotic view, life is an eminent feature of the universe itself, *and it can only be understood to the fullest possible extent if reflected from as many different ways of understanding of as many systemic contextures (and consistent conjectures) as possible*.

"The way up and down is the same," Heraclitus would have said (Kahn 2001, 74–75 [CIII]). In its totality, this new multi-valued "coin" of philosophical reflection is more than the sum of its two (co-equal) aspects (mind-centeredness and matter-centeredness at the same time). It forms the true "currency" of a poly-contextural *naturalism*—or, alternatively speaking, the true (self-)reflective *current* of poly-contextural *panenbiotism*, in which both *incomplete* sides of (the possible forms of) idealism and (the possible forms of) materialism converge in a higher philosophical understanding of *Being as an all-encompassing cosmic becoming of Life*. This "*Be-coming*" can never be reduced to *one* particular truth. It is an open system or network of singularities which are reflected as incomplete truth-like contextures.

I would like to add that the present metaphorical use of the words 'coin' and 'currency' is inspired by Heraclitus. He refers to 'fire' (including that which we today understand as the electricity that manifests itself in the form of the thunderbolt) as the universal "*currency*" of the cosmos. It represents the mediating, transforming, and connecting element between the λόγος (*logos*) and the material fabric of what Heraclitus refers to as 'air,' 'water,' and 'earth.' In a sense,

Heraclitus's world view foreshadows a panenbiotic *synthesis* of idealist and organicist-materialist perspectives beyond any one-sidedness. If we develop our thoughts in this direction and in a modern sense of a trans-classical vision in philosophy and science, it will provide inexhaustible possibilities for the production of new insights. Moreover, transcultural and trans-classical panenbiotism calls for an advanced integration of Western and non-Western traditions of philosophy *in the form of an evolving new form of poly-contextural process philosophy.*

'Cosmic Life' from a Transcultural Point of View

Against the background of what has been said so far, it is time to reflect on the term 'cosmic life' in the sense of a *comparative category*, that is, as a category (in the sense of the German term *Reflexionskategorie*) to bring together comparable views and perspectives from different times and diverse cultural contexts. In this regard, the use of the expression 'cosmic life' pertains to the reflection of a *permeable problem horizon* (Bartosch 2015, 15–17) in view of many philosophical traditions on a global scale. It is "permeable" in the sense that it allows for systematic comparative analysis of different contextures in the global history of philosophies. Based on this, we can study how comparable problems or topics have been dealt with in different and even unconnected traditions of thought and under the respective backgrounds of different histories of philosophy and the respective conceptual reservoirs and linguistic contexts, etc. With regard to the methodological foundations of transcultural comparison or the topic of comparative categories, I would like to call attention to prior works in this direction (Neville 2009, 37–38; Bartosch 2015, 18–19, 2019, 2021, 2022).

The general thesis behind the present approach can already be traced in philosophical writings of several different ancient civilizations. The first rays of this thematic horizon precede the introduction of the term *kosmisches Leben* (cosmic life) by Bronn (1841) for several thousand years. And Bronn's later concept is not to be confused with the present comparative category 'cosmic life,' which is more receptive in view of various possible contents and, thus, characterized by a sort of intended vagueness, namely for methodological reasons.) The earliest 'pre-classical' (see further above) philosophies in East Asia, South Asia, and Western Asia already heralded first elements of the panenbiotic worldview in various ways. To advance the discourse on cosmic life from classical science to trans-classical science in the sense of Gotthard Günther, these philosophies have to be explored first. Useful starting points for the further reflection on cosmic life have to be compiled and organized in view of particular subtopics and shared ideas. A basic reflection on these foundations in a great variety of ancient sources will provide valuable backgrounds for the new direction of poly-contextural panenbiotism in the global history of philosoph*ies*.

One can think of the following main segments of this chapter as a transcultural and poly-contextural introduction to a *history of the idea* of cosmic life (in the sense of an extended and comparative approach in the history of ideas). The reader will learn that the subject matter is a general one here. The *idea of life* in the sense of cosmic or universal proportions has appeared in various forms and in many cultures from earliest times onwards and in many comparable manifestations of thought. It has emerged on several occasions and independently in the early histories of Chinese, Indian, Iranian, and Greek thought (and maybe others). In addition, later European thinkers have advanced toward the problem horizon of cosmic life in the context of the further history of philosophy and science in the West. In view of these different historical contextures, the knowledge interest of the present text is motivated by the question in which way or from which contentual angle the philosophical problem of cosmic life has been addressed in each case.

It will be shown that at least *three very basic thematic fields* can be distinguished in this context. First, several traditions have reflected on the problem of cosmic life in the general sense of an evolving self(re)creation by transformative self-consumption and regeneration. In this sense, from an overall point of view, life *as such* is constantly consuming and "recycling" parts of itself, namely, to be able to (re)create more of itself—and *thus to constantly and continuously evolve in a sustainable manner*. Secondly, in various contexts, the development of ideas of cosmic life is related to reflections on (cosmic) infinity. Notions of an infinitely self-creative universe to be viewed as an infinite continuum of celestial bodies and various forms of organic life (ad infinitum) have started to evolve in various contexts. Thirdly, some philosophies discuss the topic of cosmic life as part of the more complex and more encompassing phenomenon of cosmic (natural) light—including human consciousness and thought processes as derived manifestations of (cosmic) light.

With regard to these *three major thematic fields*, we will see that (at least from the perspective of a poly-contextural and transcultural analysis) the general *idea complex* of cosmic life *is much older than its name*. Its concept-historical foundations are interwoven with the beginnings of advanced philosophical thinking in *several* ancient Eurasian civilizations—at least since the period of the "Axis Age" (Jaspers 1965) approximately 2500 years ago. Besides these layers in the global history of thought, I will also refer to a few more positions of later times as well. The aim is to provide a *prospection* of important historical substrata and thus to prepare their further integration with modern discourses, namely, to advance the transcultural poly-contextural foundations *of a future philosophical, trans-classical science of cosmic life*.

Cosmic Life as an Unceasing Transformative Self(re)creation

Thales of Miletus

The question 'What is life?' is one of the most important questions of philosophy. It is also one of the oldest ones. For example, Thales of Miletus (c.620–c.548/545

BCE), who counts as the first Greek philosopher, defines 'water' as the first *material* origin of all things. To him, however, this principle represents a *living* universe. Thales views it as the creative expression of an uncreated, *divine principle*. Moreover, it should not be forgotten that he views everything as being *animated by spiritual principles* (Diogenes Laertius 1921, 13 [I.27], 17 [I.35]). It has to be noted that his philosophy can neither be properly labeled as 'materialist' nor 'idealist.' The two camps did not exist at that time, and, in hindsight, Thales's views cover aspects of both, one might say. For this reason, Thales might even be thought of as an ancient forerunner of what has been called *poly-contextural panenbiotism* in the previous section of this investigation. At least implicitly, the directions of, so to speak, "proto-materialist" and "proto-idealist" aspects coexist in a sort of *coordinated unity through difference* in Thales's philosophy, that is, as a functional whole of different (implicit) contextures of reflection.

A basic foundation of the new poly-contextural, trans-classical approach (to be developed with regard to a new and more extended understanding of cosmic life) is implicitly "pre-formed" in Thales's ancient *pre*-classical aphorisms. One should recall the third quote by Gotthard Günther in the subsection "Poly-contexturality in the Theoretical Exploration of Cosmic Life" when reading Thales's following family-resemblant ancient remark: "Death, [Thales] said, is not different from life. 'So why don't you die,' someone said in response. He replied, 'Because it makes no difference'" (Diogenes Laertius 1921, I.35, p.17). For the classical (mono-contextural) tradition there is a complete break between life and death, between animate and inanimate matter. From a poly-contextural point of view there is not. Furthermore, this position is open for discussion also from John B. Cobb, Jr.'s process philosophical perspective presented in this volume. There is a strong conceptual "resonance" in regard to Cobb's concepts of 'pre-life,' 'living,' and 'Life Itself.' *I would like to emphasize that this correlation bears enormous potential for future reflections.*

Uddalaka Aruni, with Comparative Remarks on Other Thinkers

At this point, we have to turn to another historical context(ure) of ancient thought. Very likely, its roots are even older than those of the fragments ascribed to Thales. From the angle of another civilizational horizon, it allows to explore further what could already to be sensed in Thales's statements—namely, the idea of cosmic life as transformative self(re)creation. Uddalaka Aruni most probably lived during the 8th or 7th centuries BCE. He either lived before Thales, or he was an older contemporary. Scharfstein (1998, 55–61) introduces this ancient Vedic thinker as humanity's first documented philosopher (in the strict(er) sense). The ancient *Chandogya Upanishad* contains a dialogue between Uddalaka and his son Svetaku. As in the case of Thales, Uddalaka's ideas can be interpreted as foreshadowing a sort of *panenbiotic* position—albeit in a pre-classical or *non*-classical ("not-yet-(trans-)classical") sense. We will see that Uddalaka already nurtured a sort of rudimentary and more or less implicit concept of cosmic life (*mutatis mutandis*, in the sense of his own theory or context(ure) of reflection). This is also to say that general concepts or ideas can appear without a singular, specific expression or designation attached to it. (In analogy, one can think of peoples who did/do not have a specific term for 'mathematics' but have an implicit concept for what they are practicing in this sense, namely, in the more indirect form of a respective *set* of terms which form a system of concepts that "embodies" the general meaning of this mental activity.)

In the sense of an implicit concept(uality) of cosmic life, the ancient thinker describes *three core aspects of life* as a *material* basis of a self-conscious universe. (I am using the word 'material' as a transcultural comparative category.) These are: (1) 'warmth' (or, in modern terminology, energy-radiating processes), (2) 'moisture' and 'water' (or more generally speaking: liquids, deriving from the former), and (3) metabolic processes. The latter involve both energetic processes and liquid carriers that provide nourishment. This is to say that the earliest known philosopher of world history had already discovered that without 'warmth,' 'moisture,' and 'nourishment'—and without the related *metabolic self-transformation of life as such*—our human thinking, that is, *human self-consciousness of a universe, would be impossible*. Uddalaka presents breathing, drinking, and the digestion of food as preconditions of consciousness: "Mind, my son, comes from food, breath from water, speech from fire" (Uddalaka quoted in Müller 1879b, 98, tr. Max Müller). In regard to breath, Uddalaka also adds:

> As a bird when tied by a string flies first in every direction, and finding no rest anywhere, settles down at last on the very place where it is fastened, exactly in the same manner, my son, that mind (the *gîva*, or living Self in the mind [...]), after flying in every direction, and finding no rest anywhere, settles down on breath; for indeed, my son, mind is fastened to breath. (Uddalaka quoted in Müller 1879b, 99)

In their interdependency, all of those abovementioned three factors of cosmic life (taken in the sense of Uddalaka's ancient Vedic view) are an expression of

> [...] the True (not the Truth in the abstract, but that which truly and really exists). It is the Self, i.e. the *Sat* [सत्] is what is called the Self of everything. [...] [Uddalaka therefore] tells [his son] *S*vetaku that, not only the whole world, but he too himself is that Self, that *Satya*, that *Sat*. (Müller 1879a, xxxvi, insertions in brackets DB)

This also implies that the three material aspects only form one side of Uddalaka's Vedic *panenbiotic* "coin": Uddalaka explains that not only the world but also human consciousness in its self-manifestation as individual experience, that is, the decision-making (human) *observer* [see, *mutatis mutandis*, Maturana (2003, 57–58)], *is* that universal Self. In its universality, this Self is not the mere product of the three material characteristics of the world. When Uddalaka explains the mysteries of life and the world to his son Svetaku, he ends each of the eighth to the sixteenth khandas (chapters) of the *Chandogya Upanishad* with the same statement: "It is the Self, and thou, O *S*vetaku, art it" (Uddalaka quoted in Müller 1879b, 101–105, 107–109).

In terms of the mode of *scientific observation—of which Uddalaka's dialogue with Svetaku presents the first known historical example* (Ruben 1947, 167)—this also marks the specific difference between classical and trans-classical reflection in the sense of Gotthard Günther: In classical science, the implicit application of the logical rule of *tertium non datur* in the context of the conceptualization of a dichotomy of Being and Non-Being led to an exclusion of the question of the involvement of the investigating *subjective* observer, the self-reflective perspective of whom otherwise would present a *tertium* in such processes of conceptualization (see also further below, the next indented quote). According to Gotthard Günther, German transcendental idealism tried, but was not able to provide a fully sufficient solution for this problem (Günther 1991).

Uddalaka on the other hand already *rejects the idea of nothingness* in opposition to *sat* सत् (which is "one only, *without a second*," see further above) and identifies regenerating phases of deep dreamless sleep and our waking consciousness as functions of an all-encompassing process of cosmic life (in terms of his ancient model). In Uddalaka's view, the observer (in the waking state or in deep sleep) is always *included* in the observation. In resonance with the trans-classical approach, this also relates to 'life' and 'death': Uddalaka's thoughts already hint at the possibility of a poly-contextural approach in this regard. In the absolute sense of an identity of a trans-individual Self (of everything and being present in everything) (*sat*), and "the True," the Self is that "[…] which is, was in the beginning, one only, *without a second*" (Uddalaka quoted in Müller 1879b, 93, tr. Max Müller, italics DB). Kwee Swan Liat, a great pioneer of transcultural comparative philosophy, ascribes to *sat* an explanatory *function* comparable to ancient Greek or later European concepts of Being (Kwee 1953, 86). However, and different from classical conceptions of science (most prominently characterized by Aristotelian views), the observer is always included in Uddalaka's mode of thinking—namely, as an integral center of the self-consciousness of *sat*.

In the *Chandogya Upanishad*, Uddalaka presents this *unity of unity and difference* of living, experiencing individuals and the world process as a whole—and one has to recall that it is "without a second" (therefore representing non-countable, absolute unity)—as a self-unfolding meta-consciousness in the sense of *a cosmic willing to live*. He refers to it *as the inherent self-evocative "drive" of all life*. "It thought, may I be many, may I grow forth" (Uddalaka quoted in Müller 1879b, 93, italics DB). And "all […] creatures, my son, have their root in the True, they dwell in the True, they rest in the True" (Uddalaka quoted in Müller 1879b, 100). "When a man sleeps here, […] he becomes united with the True, he is gone to his own (Self)" (Uddalaka quoted in Müller 1879b, 98–99, tr. Max Müller), and "all […] creatures, when they have become merged in the True (either in deep sleep or in death), know that they are merged in the True" (Uddalaka quoted in Müller 1879b, 101). According to Uddalaka, every living being is connected to the deepest core of all cosmic life in the dreamless state of unconscious sleep. (Later South Asian and East Asian traditions provided meditation methods to experience this state of mind.) What Max Müller translates as 'True' is not 'the Truth' in the sense of absolute (positive) knowledge, but the possibility field of all contextures springing forth from the processual triunity of the warmth, moisture, and nourishment of cosmic life.

For a thinker like Uddalaka, the *process of the cosmic whole* could neither have been one-sidedly reduced to something like the 19th-century vulgar-materialist misconception of "building block matter" (Bloch 1972, 309, tr. DB) nor to absolutely non-material, idealistic "concept-entities." Uddalaka's cosmology presents us with a sort of very ancient *panenbiotic* view. The later particularizations in the sense of either idealist or materialist mono-contextures (which, *mutatis mutandis*, are also characteristic for the later history of Indian philosophy) had not yet occurred, so to speak. In this sense, Uddalaka's pre-/non-classical view "resonates" with the advanced trans-classical considerations discussed further above. Not unlike Gotthard Günther's poly-contexturalism, already Uddalaka rejects the idea of a mere dichotomy of Being (respectively *sat*) and Nothingness (see above: "without a second").

Also Gotthard Günther criticizes (see also the subsection "Poly-contexturality in the Theoretical Exploration of Cosmic Life") such a view as follows:

> [T]he classic pattern of thinking with its concomitant mono-contextural ontology offered no place for the observer of the world or the thinking subject because it would have been absurd to assume that the subject of *cognizing* belonged in the contexture of that which was *cognized*. On the other hand, since that which was cognizable on principle constituted the possible range of world experience, *there was no place for the subject inside the world*. Thus human thought unavoidably projected a transcendent domain beyond all Being, and *Nihility served as a very convenient vehicle for such projection*. (Günther 1979, 278–279, last and second-last italics DB)

In Uddalaka's view, the universe is presented as a self-evolving living *whole of "equi-original" contextures* of (1) a universal meta-mind, (2) a universal willing, and (3) *metabolic processes*. The latter are reflected as transformation processes of 'warmth' and 'moisture.' Their function is to enable and to sustain subjective consciousness (unique finite perspectives and processes of cognition and action). In this sense, Uddalaka is presenting Svetaku (and us) with a philosophical vision of a living universe *which is participating and reflecting itself* from finite perspectives of individual consciousness (always being reunited with the infinite core during phases of deep dreamless states of sleep).

In view of this topic of metabolism and the self-perpetuation of cosmic life by self-consummation, one can also refer to Plotinus (c.204/5–c.270). He counts as probably the most important thinker of Western Eurasian late antiquity. At least en passant, I would like to mention in this context that Plotinus states that the animals'

> [...] eating of each other is necessary. These transformations from one animal to another come about because they would be unable to continue on in existence the way they are, even if no one were to kill them. And if at the time when they leave the world, they leave it in such a way that others find some use from them, why must we begrudge that? What does it matter if they are consumed to be born as other living beings? (Plotinus 2018, 267)

In a somewhat cryptic fashion and very rudimentary form, this general understanding is already foreshadowed in Thales's aphorism on (the unity of the unity and difference of) life and death quoted further above. From the perspective of world philosophy, however, it has to be noted that it was Uddalaka's category of 'nourishment' which drew the attention to this particular thematic horizon for the very first time. This general understanding of life as a "meta-metabolic" process *is of great relevance for the present-day discussion of the general topic of cosmic life* and for the development of a future trans-classical philosophy of cosmic life.

Furthermore, the relevance of this understanding is also attested by the more recent history of philosophy. The German romanticist thinker Lorenz Oken (1779–1853) emphasizes that each single living being is just to be viewed as a moment of an everlasting, *living* world (or world organism). Like Plotinus more than 1500 years earlier, Oken *mutatis mutandis* holds the view that when an individual organism perishes, this does not mean the annihilation or destruction of life itself but only *a change within the everlasting life-process of the world*, that is, a transition into new or other life. According to Oken, there is no death per se (Oken 1809–11, 2, 17–18). Manasvi Lingam and Abraham Loeb point out that, a few decennia later, Friedrich Engels (1820–1895) reintroduced the view that one of the central characteristics of life is its "*continual metabolic interchange with the natural environment*" (Friedrich Engels quoted in Lingam and Loeb 2021, 108). In view of their astrobiological study *Life in the Cosmos*, the authors then discuss the question: "Did metabolism arise first?" (2021, 108).

According to Erwin Schrödinger's (1887–1961) famous study *What is Life?* (1944), life is to be characterized as such by way of its "feed[ing] on 'negative entropy'" (Schrödinger 2013, 70, insertion in brackets DB, see also this essay's last subsection "Connecting Dots in the Light of Solar Astronomy: Attila Grandpierre").

> How does the living organism avoid decay? The obvious answer is: By eating, drinking, breathing and (in the case of plants) assimilating. The technical term is *metabolism*. The Greek word (μεταβάλλειν) means change or exchange. Exchange of what? Originally the underlying idea is, no doubt, exchange of material. (E.g. the German for metabolism is *Stoffwechsel*.) That the exchange of material should be the essential thing is absurd. Any atom of nitrogen, oxygen, sulphur, etc., is as good as any other of its kind; what could be gained by exchanging them? For a while in the past our curiosity was silenced by being told that we feed upon energy. [...] Needless to say, taken literally, this is just as absurd. For an adult organism the energy content is as stationary as the material content. Since, surely, any calorie is worth as much as any other calorie, one cannot see how a mere exchange could help. [...] Thus a living organism continually increases its entropy—or, as you may say, produces positive entropy—and thus tends to approach the dangerous state of maximum entropy, which is death. It can only keep aloof from it, i.e. alive, by continually drawing from its environment negative entropy—which is something very positive as we shall immediately see. *What an organism feeds upon is negative entropy*. Or, to put it less paradoxically, the essential thing in metabolism is that the organism succeeds in freeing itself from all the entropy it cannot help producing while alive. (Schrödinger 2013, 71, italics DB)

It has to be mentioned that Schrödinger was very well aware of the ancient Vedic tradition of philosophy. In view of Uddalaka, it should be emphasized that it was not just the topic of reincarnation but also that of somatic *regeneration* which became very important in later Hindu traditions, too (e.g., Wujastyk 2021). In relation of this context, in which Uddalaka's thoughts and positions played a very important role, Schrödinger states:

> I know, by incontrovertible direct experience, that I am directing [my body's] motions, of which I foresee the effects, that may be fateful and all-important, in which case I feel and take full responsibility for them.
>
> The only possible inference from these two facts is, I think, that I—I in the widest meaning of the word, that is to say, every conscious mind that has ever said or felt 'I'—am the person, if any, who controls the 'motion of the atoms' according to the Laws of Nature. [...] In itself, the insight is not new. The earliest records to my knowledge date back some 2,500 years or more. From the early great *Upanishads* the recognition ATMAN = BRAHMAN (the personal self equals the omnipresent, all-comprehending eternal self) was in Indian thought considered, far from being blasphemous, to represent the quintessence of deepest insight into the happenings of the world. (Schrödinger 2013, 87)

A few years after Schrödinger, Carl Friedrich von Weizsäcker refers to this topic of metabolic interchange with the following philosophical remark:

> Life continues to evolve because every living being is on the threshold of death. The individual is in this paradoxical situation: *It kills in order to live and lives in order to die*. It produces descendants *who can only live if it dies itself* and who are destined to die themselves as well. If not it and all of its ancestors would have shared the disposition of wanting to live and having to die, it would never have come into being itself. (von Weizsäcker 1948, 97–98, tr. DB)

It has to be noted that von Weizsäcker, too, was "conversing with Sanskrit scholars" and that he had been aware of the above-mentioned Vedic concept of *sat* (Drieschner 2014, 47), which is mentioned in Uddalaka's dialogue with his son Svetaku. In a systematic sense, Uddalaka's ancient insights can thus be acknowledged as the earliest point of reference of von Weizsäcker's position.

It is highly interesting to also think about this in the sense of poly-contexturality and Gotthard Günther's understanding that "[d]eath means only a gradual decrease of the discontexturality of matter" (Günther 1979, 208) again. In this sense, 'death' in regard to one contexture means 'life' in the sense of a life-enabling, life-sustaining element (Uddalaka's 'nourishment') from the perspective of another living being or the overall process of 'Life Itself' (John B. Cobb, Jr.). We have seen that Gotthard Günther has emphasized that for classical science a complete break exists between life and death. From an advanced poly-contextural point of view, on the other hand, everything is to be seen in its overall relatedness to everything else and in the sense of a certain undecidability. Uddalaka's concept of nourishment already implies that (cosmic) life is to be characterized in a certain wide sense of 'metabolic processes,' that is to say, as a higher form of consciousness-bearing, *autopoietic*, "matter-energetic" self-transformation.

Daodejing, Liji, and Wang Yangming

Apart from ancient Greece, Vedic India, or the much more recent German discourses of the 19th and 20th centuries—where else do we find traces of a *panenbiotic* understanding?

One of the ancient sources that we can turn to is the Chinese Daoist classic *Daodejing* 道德經. This work was most likely edited during the 4th and 3rd centuries BCE. Pre-modern scholarship has ascribed it to the semi-legendary thinker Laozi 老子 (also Lao-tzu), who is said to have lived during the 6th century BCE and to have been an older contemporary of Confucius. 'Laozi' (Master Lao) is an honorary name. According to traditional literature, his real name was Li Er 李耳 and his courtesy name Boyang 伯陽. As this thinker is said to have drawn thoughts from older sources, it is likely that some of the book's content might even precede the time of the 6th century BCE. In the sixth chapter of the *Daodejing*, we can find such a passage:

> The spirit of the valley does not die. It is called the mysterious femaleness. The gate of the mysterious femaleness means the root of Heaven and Earth. As if it were a (self-continuing) thread of silk, it does (always) remain, and its effectuation is effortless. (*Daodejing*, [§] 6, tr. DB)

Its antiquity is supported by its traditional attribution to the legendary Yellow Emperor (Huangdi 黃帝) (*Liezi*, n.d., [I] "Tianrui", [§] 1). According to traditional historiography, this legendary figure is supposed to have lived during the end of the first half of the third millennium BCE. The translation terms 'spirit of the valley' (*gu shen* 谷神) and 'mysterious femaleness' (*xuan pin* 玄牝) allude to an *ever-ongoing and boundless becoming* per se. In the sense of an ancient Chinese vision of cosmic life (please remember that 'cosmic life' is used as a comparative category here), this (Daoist) meaning of *dao* 道 fulfils the function to semantically allude to an infinite, all-including—yet immanent and ineffable—common "denominator" *within* the constant unfoldment of *everything* in the process(es) of existence.

This ancient Daoist contexture of the philosophical problem of cosmic life differs from that of Uddalaka in its original Vedic environment of thinking. Both views relate to independent origins in the global history of philosophies. From the perspective of poly-contexturality as well as transcultural comparative philosophy in general, both represent valid reflections under the backgrounds of their times and cultural contexts. We do not have at our command an independent observer position to judge these reflections from the knowledge perspectives of our own contemporary or cultural contexts or in the sense of evaluating one of these historical reflections against the other. However, we can say that, *mutatis mutandis*, the term *dao* as well as expressions referring to life and fertility, like *gu shen* (spirit of the valley) or *xuan pin* (mysterious femaleness) fulfil a comparable basic *explanatory function* as the term *sat* does in Uddalaka's reflections.

According to the German Buddhist Lama Anagarika Govinda (Ernst Lothar Hoffmann, 1898–1985), the universal *life principle* of the "mysterious femaleness," *dao*, means a

> [...] *world progress*, of the cyclical movement of the *living universe* which, like a river, creates law and order through the stability of its orientation and its own innate rhythm, and brings forth all outward appearances and living forms, animates them, and permeates them with meaning. (Anagarika Govinda 1992, 16, second italics DB)

In analogy to Uddalaka's ancient non-classical position, the writer(s) of *Daodejing* did not endorse the concept of absolute nothingness—they put the idea of formlessness or formless existence in its place. Instead of a *Kreationslogik* (logic of creation), that is, the idea that the creative origin of the world is—not in a temporal but *in a logical sense*—prior to the world (Bartosch 2015, 26–68, 124–156), the Daoist position is based on a sort of *Transformationslogik* (logic of transformation) (Bartosch 2015, 68–122, 156–190; Li 2000, 132–138, 294). According to this, the world process is to be reflected as a constant and boundless self-organization of an original totality of life. This totality is unfolding itself in correlated changes of creative individualization, that is, in the sense of an infinite process of immanent self-differentiation. Like in our former case (of Uddalaka), the *observing* human consciousness is thought of as an inseparable, and even indispensable, element of a cosmic process of life—which has to be thought of as all-encompassing and even including the motions of the celestial bodies.

Mutatis mutandis, this also relates to Confucian thought. In the ancient classic *Liji* 禮記, the form and the actual (inter-)subjective experience of each human being is defined as *tian-di zhi xin* 天地之心, the 'heart-mind of Heaven and Earth':

> Human—from time immemorial, this (word) means the excellence of the heart-mind of Heaven and Earth, as a crossing of *yin-yang* and as the get-together of (earthly) spirits and (heavenly) gods, the pure flow of *qi* 氣 ['fluidum' or 'matter-energy,' see also Bartosch 2015, 83–85, etc.] of the five phases (*wu xing* 五行) [see also Needham 1977, 253, etc.]. [...] This is why the human being is the heart-mind of Heaven and Earth, the foundation of the five phases. (*Liji*, n.d., "Liyun," [§] 20, tr. and additional references in brackets DB)

Later, the important Neo-Confucian philosopher Wang Yangming 王陽明 (also: Wang Shouren 王守仁, 1472–1529), who also studied Buddhist and Daoist ideas, states: "Heaven and Earth, the ten thousand things—at bottom they all form one body with me" (*tian-di wan wu ben wu yi ti* 天地萬物本吾一體) (Wang Shouren 1933a, 76, tr. DB). The basis for this understanding was that everything is carried by one and the same never-ceasing flow of 'matter energy' (Joseph Needham's translation for *qi* 氣); everything represents "one and the same *qi* in circulation" (*yi qi liu tong* 一氣流通) (Wang Shouren 1933b, 33, tr. DB). Under this background, Wang Yangming emphasizes that the 'human being' (*ren* 人) commands an 'empty spiritual force' (*xu ling* 虛靈) and has an intuitive conscience, that is, a situational *liangzhi* 良知 (knowing of goodness), which is inseparably or necessarily in touch with Heaven, Earth, and the ten thousand things at all times (Wang Shouren 1933b, 17).

From a comparative angle, it might be added that Wenchao Li points out that Neo-Confucian ideas like this "threaten to break down any boundary between spirit and matter, between *being and non-being* in [mono-contextural] Christian thought" (2000, 141, tr., insertion in brackets, and italics DB). In parallel, this, of course, can also be related to Gotthard Günther's earlier-mentioned remarks in "Life as Poly-Contexturality" and to his respective views on trans-classical science. Also in this context, a certain affinity between these layers of ancient or pre-modern non-Western thought and Günther's approach cannot be ignored.

Yijing (Book of Changes)

What is the wider background for this ancient Chinese *panenbiotic* way of thinking? And how does this relate to our question of a multifaceted appearance of the idea of cosmic life in many civilizations (and in general)? From an ancient Chinese angle, everything is to be considered as changing and shifting situations—even 'things' in the sense of categorizable objects can be understood as "standard situations" or processual constellations (so to speak) in the context of the self-perpetuating, ever-flowing, self-changing order of the whole. And, at the very core, *life has no beginning nor end: its permanent change is what reflects its constancy*.

The Austrian cultural anthropologist Ernst Viktor Zenker (1865–1946) emphasizes that, according to ancient Chinese philosophy, the creative act *that provides and sustains life takes place in the mind as well as in the body* (Zenker 1932, 46). This represents a "non-classical" panenbiotic position which is in a sort of systematic resonance with the ancient Indian ideas of Uddalaka:

> Thus, at the heart of ancient Chinese philosophy, *we find a naturalistic* [in the present terminology: panenbiotic] *and not a materialist idea. Nature is everything, not only matter, but also spirit.* God, briefly called [*Di*] or the Lord in the [*Yijing*], is [...] in nature, and the cosmos becomes cosmos, order, or, as the pseudo-Aristotle says, σύστημα [ἐξ οὐρανοῦ καὶ γῆς καὶ τῶν ἐν] τούτοις ηεριεχομένων φύσεων, only through the mind. (Zenker 1932, 46, tr., italics, and insertions in brackets DB)

This is perfectly in line with the poly-contextural perspective, which is (to advance Hegel's words) "elevating, negating, and preserving" the dichotomies of Aristotelian versus dialectical logic (also in the sense of "implicate logic," [Bartosch 2022]) as well as the contradiction of possible idealist and materialist views in the light of the new level of understanding of the *panenbiotic* world outlook and its orientation toward cosmic life.

This new level of thought is already "pre-reflected" in the "information," that is, the process-constituting and (in)forming symbolism of the ancient classic *Yijing* 易經 (*Book of Changes*). Here we find two binary symbols—a solid line (*yang* 陽) and a broken line (*yin* 陰), both of which may either symbolize a steady situation (continuous *yin* or *yang*) or a situation of change, that is, in the sense that their corresponding momentary value is about to switch to the respective opposite (an *yin*-value turning into *yang* or vice versa). According to the system of change(s) (*yi* 易), both elementary symbols can be further combined in doubles or pairs of opposites to form four basic combination patterns of *yang-yang*, *yin-yin*, *yang-yin*, and *yin-yang* (indicating more complex processes and patterns of change). Furthermore, three of those one-valued binary symbols (each of them to be identified as either *yin* or *yang* as well as either steady or turning) combined together result in eight possible combinations, including further internal (situational) differentiation (according to the status of each *yin* or *yang*). These eight trigrams allow the representation of even more complex patterns and cycles of change. Finally, all eight trigrams can be combined with each other, which means the same as the combination of six of the basic one-valued symbols (either *yin* [either steady or changing] or *yang* [either steady or changing]) in the form of sixty-four basic hexagrams (for more, see also Hertzer 1996 and Nielsen 2003).

In the sense of this very ancient Chinese *information* system, the binary code with its value ascription of steady or changing states forms the basis to represent immanent relationships of life, the biotic relations that form ecosystems, processes of experience, internal (metabolic) processes of living organisms, courses of situations, etc. It shall be mentioned that this universal code of cosmic (ex)change was known to Gottfried Wilhelm Leibniz (1648–1714), who introduced the binary code and is considered to be one of the very early pioneers of mechanical computing (Davis 2018, 1–14). The Swiss psychologist Carl Gustav Jung (1875–1961) emphasizes that the combinatory symbolism of the *Yijing* can be related to his theory of the collective unconscious as well as to the phenomenon of synchronicity (Zhang and Shen 2021). In an even more astonishing parallel, Martin Schönberger reveals how the *Yijing* represents *the same formal systematics* as the DNA code discovered by Watson and Crick in 1953 (Schönberger 1992).

Zoroastrian Views and Heraclitus of Ephesus

The ancient Persian Zoroastrians developed a "pro-cosmic dualism" (Boyce 1996, 230) and particular ritualistic forms of life and social interaction, thereby nurturing the view that all life is cosmic, that is, the all-encompassing, the central element in the creation-process of everything. Certain similarities to Uddalaka's and other ancient Vedic views can be noted: All life is deriving from the light and warmth of the sun that keeps everything alive. This is symbolized by the holy fires in Zoroastrian temples. 'Fire'—or more generally speaking and in modern terminology: the release of energy—is the main attribute of the *good* creator Ahura Mazda.

It has to be noted that "the world [of the Zoroastrian] is [originally and inherently] good and evil attacks it from the outside" (Boyce 1996, 230, insertions in brackets DB). This evil from the outside attacks humans in the form of selfish desires and greed. *It attacks the human being from beyond its humanity*. According to Zoroastrian lore, these occurrences of evil have been disturbing the moral and *ecological* integrity of humanity in the form of evil spirit(s) since the beginning (*Der Bundehesh* 1868, 19–20). However, Zoroastrians view the historical return of humanity to the good life and to a primordial goodness of all life as *an undeniable eschatological promise*. From their perspective, this element of hope even represents the general meaning of life as such. Against this background, one might also conjecture that this ancient belief, either directly or indirectly, might have played an eminent role as an original source for many of the later utopian visions of Western Eurasian philosophers, for example, in the Mediterranean, Greek or Latin-speaking regions, or in later European contexts as well.

It is possible that Heraclitus of Ephesus was inspired by the Zoroastrian teachings of his Persian neighbors. As mentioned, he assumes that what he refers to as 'fire,' including the thunderbolt, represents the cosmic "currency" and thus the root of all life. Everything comes from 'fire,' respectively, from a dispersion and transformation of *energy*—and everything has to return to it. "The ordering [κόσμος], the same for all, no god or man has made, but it ever was and is and will be: fire *ever-living*, kindled in measures and in measures going out" (Heraclitus, fragment XXXVII [D. 30, M. 51] quoted in Kahn 2001, 45).

Flickering flames and fire-igniting thunderbolts were reflected as a manifest synthesis of the principle of the λόγος (*logos*), that is, the (*implicit*) *order* of the cosmos, but also of human speech and reason—and, on the other side, the first principle in the interflow and transformation of the basic states of nature; besides 'fire' these are 'water' (fluid states of matter), 'air' (gaseous states), and 'earth' (solid states). Heraclitus's understanding of *logos* is not unlike that of *dao* in ancient Chinese thought (Wohlfart 2001, 165–178). It reflects a self-realizing form of an ever-ongoing, self-creating process of reciprocal alternation as a dynamic unity of opposites in the sense of the dialectical identity of day and night, waking and sleeping, of birth and death, beauty and ugliness, etc. It should be noted that Heraclitus inspired Georg Wilhelm Friedrich Hegel's (1770–1831) central idea of *Being as dialectical Becoming*, that is, as a self-unfolding process of progressing steps of contradictions toward the absolute intension and infinite reality of the idea (Hegel 1986, 84). Hegel's approach, in turn, became a starting point for Gotthard Günther, who states that "the new trans-classical [poly-contextural] rationality [already] tries to have its say in Hegel's work" (Günther 2005, 3, tr. and insertions in brackets DB).

In Persian Zoroastrian mythology, in the "Av[estan language, the first human being is called] *Gayō.marətan*, 'Mortal Life'" (Boyce 1996, 97, insertion in brackets and italics DB). In ancient Greece, humans were defined as mortals in contrast to the immortal gods of the Hellenic pantheon. (The latter were believed to determine and influence the lives of the 'mortals.') Heraclitus of Ephesus also reflects a lot on the human condition of mortality. It is precisely in this context that he develops his abovementioned (ancient) dialectical views (which have become so influential since the 19th century): "Immortals are mortal, mortals immortal, living the others' death, dead in the others' life" (Heraclitus, fragment XCII [D. 62, M. 47] quoted in Kahn 2001, 71). Again, we find the idea of *cosmic life as an unceasing transformative self(re)creation.*

Heraclitus influenced Hegel in this regard as well. Hegel said that Heraclitus's remarks on the dialectical *identity of life and death* express the unity of Being and Nothingness (Hegel 1986, 84). This, of course, means just another touch point with Günther's poly-contextural understanding of life. Furthermore, Martin Heidegger (1889–1976) was much inspired by Heraclitus's reflections on mortality (Heidegger 1967, §§ 46–53). Last but not least, this also presents us with another "string of thoughts" that resonates with the views of Thales, Uddalaka, Plotinus, Lorenz Oken, the philosopher–scientist Carl Friedrich von Weizsäcker, etc. that were mentioned further above in the preceding subsections.

That which has to die forms an indispensable unity with that which cannot die, that is, with cosmic principles (in Heraclitus's terminology: *logos* and the transformations of the elements deriving from and geared toward what he means by 'fire'): "The opposites are one; and this *deathless structure* of life-and-death [of the κόσμος] is deity itself" (Kahn 2001, 23, insertion in brackets and italics DB). It is the self-appearing *logos* that "steers all things through all" (Heraclitus, fragment LIV [D. 41, M. 85] quoted in Kahn 2001, 55), that is, *the life of the κόσμος itself.*

In this sense, Heraclitus defines *immortal* cosmic life as a "deathless" relationship of an infinite actualization of interconnected individual lives and their interconnected deaths (in the sense of constant transformation in the form of new manifestations of life). In a cryptical fashion, Heraclitus interprets the "self-conflicting" nature of cosmic life as a tense harmony and as a unity of diverging opposites: "The name of the bow is life; its work is death" (Heraclitus, fragment LXXIX [D. 48, M. 39] quoted in Kahn 2001, 65). The ancient Greek words for 'life' (βίος *bíos*) and 'bow' (βιός *biós*) are basically the same. They only have different accent marks—a feature of the Greek written language that was introduced centuries after Heraclitus's time. Heraclitus views life as a state of more or less tensed dynamic equilibria of diverging yet unified forces which form one cosmic process. Heraclitus's aphorism partly implies the investment of energy (the metaphorical "stretching of the bow") to absorb other processes of (cosmic) life, namely also to kill for survival, and it also implies to be killed or to die so that other processes of the cosmic nexus of life can be sustained.

Interim Summary

There is one basic idea that is somehow "shimmering through" all of the South Asian, East Asian, West Asian, and European sources that have been discussed here so far. Comparatively speaking, *life as a whole* (in the sense of those diverse *panenbiotic* outlooks that have emerged in various civilizational contexts and at particular times) *is constantly killing-and-reviving itself, restructuring itself, rearranging itself, reallocating itself. Cosmic life—life in its totality and all-embeddedness—invests or sacrifices states of itself to recreate itself over and over again in higher and more evolved forms of self-expression; cosmic life provides certain aspects of itself to itself as its own nourishment, especially in the form of light* (see also further below in the section "Cosmic Life and Cosmic Light"), *and to produce more of itself, to grow and expand as much as possible in ways that are favorable for itself as a whole.*

Cosmic Life and the Question of Its Infinite Extension and Distribution

Numbers as an Expression of the Infinite in Daoist and Pythagorean Philosophy

Both Laozi and the Hellenic philosopher Pythagoras (both 6th century BCE, according to pre-modern sources) have been attributed with the insight that all things in the world presuppose a ternary number-ontological foundation in the sense of a most basic trinary principle as the basis of all thinking and everything there is. It is represented by the natural numbers one, two, and three. In the later Hellenistic, Roman, medieval, and early modern contexts, it was discussed under the term μονάς or *monas* (Lötzsch 1984, 114–17), which also prepared Leibniz's conceptualization of the monad.

Laozi says: "*Dao* brings forth (the number) one. The (number) one generates (the number) two. The (number) two generates (the number) three. The (number) three generates all things" (*Daodejing*, [§] 42, tr. DB). *Dao* is the unlimited, the unnamable (*Daodejing*, [§] 1). It contains everything in the sense that it represents the *non-definable* ("in-*finite*") origin-totality of all developments and is simultaneously effective in every single thing. The number one, which is the most original measure, is *dao*'s self-embedded "means" to logically "translate" itself/everything into existence (toward the multiplicity). Therefore, the number 'one' implicitly symbolizes the concept of (infinite) unity, and it also represents the most basic (finite) unit *at the same time.* In other words: it implies one*ness* or unity (non-countable), and it is the most basic unit (countable) *at the same time*. It has an intellective and a rational aspect, because it symbolizes the unity of unity and difference. The *conditio sine qua non* of every possible difference (and therefore of every possible cognition)—which implies at least a differentiation of two units (a "two-ness")—is the unit-character which is represented by the number one. In a likewise fashion, "Pythagoras treated the One as equal to the unit, the perfect and primary number, which is the essence of all creation and at the same time the creator of the material unity of things, and, of course the sole principle of all other beings" (Theodossiou et al. 2010/2011, 168; see also Aristotle 1956, 5).

By means of an *immanent* self-differentiation of the former concept of "*one*-ness", we receive the concept of duality ("two-ness")—as it is represented by the number two. At the same time, the number two thus represents the unity or synthesis of two *different* "one-units." The character of the *unity* of the number two is *not* the same as the inherent unity of the number one, that is, it is not to be viewed as *the* unity of infinite unity and the concept of finite singularity. The inherent unity of the number two (as a "two-unit") is to be viewed as a *unity through the difference* of the divergent partitions (two singularities) of the former concept of "oneness" (represented by the number one).

How is the (number) two brought forth from the (number) one, as Laozi said? The one "begets" the two in that the *emphasis shifts* from the *unity* of the unity *and* difference (of infinite unity and its finite representation as a singularity-unit), which is symbolized by the number one, to the *difference* in the unity of unity and difference (of two distinct singularity-unit*s*). Put differently: infinite unity can only *manifest* itself (a) in correlation with and (b) in contrast to a finite "image," that is, the wholeness of a most basic *singular unit: the number one* (per se). It is exactly this implicit finiteness of the representation of singularity (unit-*ness* instead of unbound unity) which bears the *possibility* of *activating the difference* in that singularity. This is how polarity, the unity of *two* (singular) units, the number two—duality—is "born."

The implicit idea of *duality* and "two-ness" then provides the logical foundation (a priori) to fuse, that is, to add up *both* (unequal!) unit*ies*—the number one (non-differentiated as a unit/singularity) and the number two (differentiated, divisible into two one-units)—in the sense of a further *unity of their forms of inherent unity and difference*. This, as we all know, accounts for *three* "one-units," respectively, *two* unequal unit*ies* as concerns the distribution of "one-units." The number three is "descending" from the "two-unity," which also inherits the singularity of the "one-unit," in the sense that *both* can only be added up, because *we differentiate between them.*

Those numbers which can be seen as the synthesis of two identical unities (consisting of the same amount of one-units each, that is, starting from the idea of an even "two-ness") we call even numbers; those which are the quantitative sum of the one-units of two *successive* and therefore non-identical positive natural numbers (respectively unit*ies* with an unbalanced "two-ness", differing by the amount of one "one-unit") we call uneven or odd natural numbers. Based on this logic of the self-differentiation and self-unfoldment of Laozi's *dao* 道 and the (same!) infinite Pythagorean number-origin of a non-countable yet difference-implying *unity* (as such), which, being the inherent foundation of the countable *unit* '1' *at the same time* (see also the next subsection), leads to '2' and from both to '3'—all following natural numbers can be derived and developed. This is why Laozi says that the number three brings forth all things. In addition, we should not forget that the sequence '1–2–3' is also the starting progression of the Fibonacci Sequence, which, as we know today, can be related to the structure of organisms and processes like the formation of galaxies, etc.

The Idea of an Infinite Distribution of Life in Hellenic, Roman, and Chinese Thought

It is remarkable that both semi-legendary figures Laozi and Pythagoras (in the sense of a most basic self-reflection) are said to have traced back the fundamental condition of all numbering, counting, measuring (and therefore also cognition in general) to an infinite unity which contains all *possible difference*—or rather, the overall-possibility of cosmic events of differentiation.

Moreover, "[a]ccording to Pythagoras itself the world was boundless" (Theodossiou et al. 2010/2011, 167). The respective Greek term—ἄπειρον (the infinite)—became most prominent in the philosophy of Anaximander of Miletus (c.610–c.546 BCE), but it was also used by others, like, for example, Pythagoras or Democritus (461–371 BCE). Therefore, it is not surprising that the idea of an infinite cosmos and an infinite plurality of worlds took root in Greek thought. According to Diogenes of Apollonia (5th century BCE), "*[t]here are worlds unlimited in number, and unlimited empty space*" (Theodossiou et al. 2010/2011, 167, italics DB; see also Diogenes Laertius 1959, 469). Simplicius of Cicilia (c.480–560) reports: "Leucippus [5th century BCE] and Democritus say that there are infinite Worlds inside the infinite void, which were formed from an infinite number of atoms" (Theodossiou et al. 2010/2011, 169; see also Simplicius 2014, 18 [202]).

Laozi's abovementioned reflections on the world-process(es) ("the ten thousand things" [*wan wu* 萬物]) as a numerical self-unfoldment of an infinite (and ineffable) foundation ("*dao* 道") is inseparably related to the comprehensive system and symbolism of the ancient information system of the *Yijing* (*Book of Changes*). In addition to the fact that something like *the idea of inert matter in the sense of "dead substrate" is absent in this context*, the numerical progressions and transformations associated with the workings of the *Yijing* were understood as the *self-organization of the life of the world as a whole* (see also the subsection above, "Yijing (Book of Changes)," and also the remark on the work of Martin Schönberger there).

The ancient Chinese word for the cosmos (in present-day standard pronunciation and writing) is *yuzhou* 宇宙. It literally means 'space–time' and has been introduced by the legalist thinker Shi Jiao 尸佼 (c.390–330 BCE): "The four cardinal directions as well as above and below are called space. And the earliest past until the present day is called time" ("上下四方曰宇, 往古来今曰宙" [Shizi, n.d., tr. DB]). Legalism (*fajia* 法家) is one of the major schools of ancient Chinese philosophy. Regarding its origins, Legalism is connected to Daoist thinking. Since the Han dynasty (202–9 BCE and 25–220 CE), it has also influenced the development of Confucian state doctrines. The *Yijing* was revered by most ancient schools of Chinese philosophy, even by the Legalists of the Qin dynasty (221–206 BCE), who were very intolerant otherwise.

Zhang Heng 张衡 (78–139), the most important polymath and inventor of the Eastern Han dynasty, later added the following to Shi Jiao's definition of *yuzhou* (space–time): "The exterior of space is boundless. The moment of time is infinite" ("宇之表里无极, 宙之端无穷" [Zhang Heng, n.d., tr. DB]). Zhang Heng distinguishes between observable space and an infinite circumference of the whole. Although his views then were mostly neglected by his contemporaries, traces of the idea seem to have inspired later Chinese authors on occasion.

More than 1000 years after Zhang Heng, Deng Mu 邓牧 (1246–1306) finally came up with the *idea of a multiplicity of inhabited worlds*:

> Heaven and Earth are large, yet in the void (of space) they are nothing more as a single grain in the void. The void (space) is the root, and Heaven and Earth are like the fruits (of a tree); the void is like a kingdom, and Heaven and Earth are like single human beings. What is brought forth from one root (of a fruit tree) cannot be (represented by just) one single fruit; when a kingdom is brought into existence this cannot be (the work of just) one single human being. Therefore, it is not reasonable to assume that there are no other structures like Heaven and Earth outside of (our) Heaven and Earth. (Deng Mu, n.d., tr. DB)

From a systematic (or comparative) transcultural perspective, it is worthwhile to glance back at the developments in ancient Greek and later Roman contexts from here. It is remarkable that *both* ancient Greek idealist and materialist thinkers have developed their own respective views in regard to cosmic life. Plato clearly defines the cosmos (as he viewed it) as a living being (albeit not in the sense of a multiplicity of inhabited worlds). The aforementioned atomistic materialists Leucippus and Democritus explicitly combined the views that (a) space is infinite and that it (b) contains an infinite number of atoms with the understanding that (c) life (atom-based, in Leucippus's or Democritus's ancient sense) is a ubiquitous characteristic of the whole.

Epicurus (341–270 BCE), who firmly believed in the ubiquitous proliferation of life throughout the cosmos (Crowe 2008, 4–5), writes:

> Furthermore, there are infinite worlds both like and unlike this world of ours. For the atoms being infinite in number, as was proved already, are borne on far out into space. For those atoms, which are of such nature that a world could be created out of them or made by them, have not been used up either on one world or on a limited number of worlds, nor again on all the worlds which are alike, or on those which are different from these. So that there nowhere exists an obstacle to the infinite number of the worlds. (Epicurus 1926, 25)

Chandra Wickramasinghe recently referred to an Epicurean thinker, who argues: "It is unnatural in a large field to have only one shaft of wheat and in the infinite universe only one living world" (Metrodorus the Younger of Lampsacus [c.331–c.278 BCE] quoted in Wickramasinghe 2021; see also Crowe 2008, 5). The Epicureans were also influenced by Democritus's atomist world view. One also has to note how close this is to Deng Mu's aforementioned fruit metaphor.

Chris Impey reminds us that Aristarchus of Samos (c.310–c.230 BCE)—the first known thinker to come up with a (scientific) heliocentric world view—envisioned a "universe [that] was one billion miles across, a phenomenal number in an age when most people never ventured far from where they were born" (Impey 2007, 7, insertion in brackets DB).

> Archimedes [c.287–212 BCE] even used the Aristarchan Sun-centered model to estimate the amount of matter in the universe. [... H]e estimates that the universe is several trillion miles across and calculates that it would take a staggering 10^{64} grains of sand to fill it. If we imagine that these grains are clumped into planets and spread over the much larger volume of the modern universe, we can calculate the number of Earths it would contain: 10^{33}, a billion trillion trillion. (Impey 2007, 8, insertions in brackets DB)

In his important study of the subject, Steven J. Dick states that "[t]he birth of the idea of other worlds in its earliest form is almost coincident with the birth of Western science in ancient Greece" (Dick 1982, 6). Furthermore, Dick draws attention to the fact that the aforementioned Epicurus therefore somehow is to be seen as an antipode to his older contemporary Aristotle (384–322 BCE) (1982, 6), who was not only partly responsible for the rejection of Aristarchus's sun-centered cosmology (Impey 2007, 7–8), but also for the establishment of the position that "there cannot even be more than one world" (Aristotle 1939, 69) as well as that "the world must be unique [and that t]here cannot be several worlds" (Aristotle 1939, 75, insertion in brackets DB) and, finally, that the "centre of the cosmos, which is unmoved and fixed, is occupied by one exclusively 'life-bearing earth,' the home and mother of living beings of all kinds" (Aristotle 1955, 347, 349). Further remarks in the *Metaphysics* show that "Aristotle's position concerning the question of other worlds was linked not only to his physics and cosmology, but also to his theology" (Crowe 2008, 13). To my knowledge, Stephen J. Dick was the first to show in great detail that this dichotomy of many-worlds theory and one-world theory has been a driving element in the history of European science from "Democritus, Epicurus, Lucretius, Aristotle, Thomas Aquinas, William of Ockham, Nicole Oresme, Johannes Kepler, John Wilkins, Bernard le Bovier de Fontenelle, Christiaan Huygens, and Immanuel Kant" (Dick 1982, 3).

This also deserves some consideration in the sense that the many-world theories can be viewed as implicit initiations or emergences of *trans-classical science* and of possible *poly-contextural* understandings of the cosmos: Every world is unique and is uniquely embedded in its own unique systemic context (see also the following subsections on Cusanus and Montaigne). Each of these world systems is consistent in itself—but all of them form one infinite cosmic nexus (ad infinitum). From this angle, one might say that Aristotle's approach is a sort of prime example for the *mono-contextural constriction* that even today is still hindering progress, or rather, the acceptance of alternative views and attempts to scientifically explore the question of the origin(s) of life in the sense of a cosmic foundation itself—or at least as a cosmic phenomenon in the more extended sense (see also the chapter by Wickramasinghe in this volume). In all fairness, it must also be mentioned again that Plato had already rejected the many-worlds theory. Gotthard Günther viewed Plato's *mono-contextural* epistemology of an exceptional and absolute truth, that is, all possible knowledge (allegedly) leading to so-called absolute "truth" in a systematic pyramid-like fashion, as the major function of classical science. From here, one could also ask if one-world theories or "exceptionalist" views in regard to the possibilities and origins of life represent a (misleading) logical consequence of mono-contexturality in thinking.

Furthermore, it is interesting to note that many-world theories in relation to the idea of an infinite cosmos were also proposed in the centuries after Aristotle. Titus Lucretius Carus (94–49 BCE) proclaims: "Nothing in the universe is unique and alone, and therefore in other regions there must be other Earths inhabited by different tribes of men and breeds of beast" (Lucretius quoted in Wickramasinghe 2021). Lucretius also emphasizes that "*nature is free and uncontrolled by proud masters* and runs the [infinite!] universe by herself without the aid of gods. For who [...] can rule the sum total of the *measureless*?" (Lucretius quoted in Crowe 2008, 7–8, insertion in brackets and second italics DB) Again, we have found a forerunner of the panenbiotic worldview here (the potential for further transcultural comparison included).

In *Refutation of All Heresies*, a work which was formerly ascribed to the Roman Church father Hippolytus (170–236), reference is made to Democritus's assumption that livable planets and life are a common feature of an infinite cosmos:

> Worlds are infinite in number and vary in shape. In some of them, nor [sic] Sun nor Moon exist, others have bigger and others have more Suns and more Moons. The distances between the worlds are unequal and in this part of the void may there be more worlds and in another fewer. And other worlds still developing, and others reached the peak and others decay, others are born and others disappear. The worlds are destroyed when they collide. There are also worlds bare of plants and animals and with no water. (Hippolytus quoted in Theodossiou et al. 2010/2011, 169; see also *Refutation of All Heresies* 2016, 47)

Nicolaus de Cusa on the Idea of an Infinite Distribution of Extraterrestrial Life

Approximately at the same time as the abovementioned Deng Mu in China, the idea of a plurality of worlds started to gain momentum in Europe again as well. While during the High Middle Ages, the mono-contextural Aristotelian dogma of the one-world theory had prevailed at first, the tides finally began to turn during the last quarter of the 13th century (Dick 1982, 23–28):

> At Paris Godfrey of Fontaine, Henry of Ghent, and Richard of Middleton held that a plurality of worlds was not theologically impossible; at Oxford William of Ware, Jean of Bassols, and Thomas of Strasbourg made the same claim. [...] They were also the first to hold that God could create other worlds, and that every possible world could determine the natural motions of its own bodies, whether those bodies were different in form from those of our world or the same. (Dick 1982, 28)

Another 150 years later, the understanding of an *infinite* universe, *including or manifesting an infinity of worlds*, started to gain momentum in occidental contexts again. For Nicolaus de Cusa (1401–1464), an omnipotent, infinite creator implied its own 'manifestation' (*apparitio*), that is, its infinite image or "projection" in the form an infinity of worlds (Goerwitz 1985, 423). In *De docta ignorantia*, the famous Renaissance thinker emphasized his conviction that life is *a ubiquitous feature of an infinite cosmos*. This view was "totally antithetical to Aristotle" (Dick 1982, 41).

In the context of the "implicate-logical" foundation (Bartosch 2022; see also further above, "Extending Classical Science Toward Transcultural and Transclassical Views") of his notion of *coincidentia oppositorum*, Cusanus views everything as a unique expression of an *absolute* synthesis of the infinitely smallest and the infinitely largest. In its own way, everything represents an *immanent space–time-transcendent oneness*. It has to be added that Cusanus's related way of thinking gave rise to many new ideas which later turned out to be valid and of great inspirational value to others. For example, Cusanus's view that our planet is not a perfect sphere but slightly flattened at the poles inspired Johannes Kepler's (1571–1630) understanding of elliptical planetary orbits. In Cusanus's sense of the related intellective vision of a higher form of *docta ignoratia* (knowing non-knowing; German: *wissendes Nichtwissen* [Mojsisch 2001, 240; Bartosch 2015]), every being and every point in the universe is—to use one of Cusanus's own philosophical *aenigmata*—both at the center and at the circumference *at the same time*:

> For although God is the center and circumference of all stellar regions and although natures of different nobility proceed from Him and inhabit each region (lest so many places in the heavens and on the stars be empty and lest only the earth—presumably among the lesser things—be inhabited), nevertheless with regard to the intellectual natures a nobler and more perfect nature cannot, it seems, be given (even if there are inhabitants of another kind on other stars) than the intellectual nature which dwells both here on earth and in its own region. For man does not desire a different nature but only to be perfected in his own nature. (Nicholas of Cusa 2001, 96, tr. Jasper Hopkins)

The superelevation of human nature in comparison to possible "inhabitants of another kind on other stars" in this passage might be a concession to the *zeitgeist* (religious dogmas) of the era. In my view, when Cusanus speaks of a "mediating human nature" (*natura media humana*) (Nicolaus de Cusa 1932, 144) and defines the human being as a "secondary god" (*secundus deus*) (Nicolaus de Cusa 1988, 7), this is more to be taken as a *case example* for the status of all beings which intellectively participate in the (self-)reflection of the divine mind (*mens divina*) on a cosmic scale.

This can also be derived from the fact that Cusanus sort of purposefully self-contradicts the above-stated anthropocentric opening in the following passages of the *same* text. He thereby expressed his idea that *the universe forms an infinite organism of cosmic life*, which is to be viewed as an infinite image of an eternal, boundless creator "who" thereby reflects on "himself" by way of "his" projections, that is, finite (living) minds being capable of reflecting back on themselves and the inherent principle of this 'self-reflection'. As these represent an infinite number of (finite) perspectives, the inhabited regions of the cosmos therefore also have to be infinite, according to the cardinal's understanding:

> [T]he inhabitants of other stars—of whatever sort these inhabitants might be—bear no comparative relationship to the inhabitants of the earth (*istius mundi*). [This is true] even if, with respect to the aim of the universe, that entire region bears to this entire region a certain *comparative relationship which is hidden to us*—so that in this way the inhabitants of this earth or region bear, through the medium of the whole region, a certain mutual relationship to those other inhabitants. (By comparison, the particular parts of the fingers of a hand bear, through the medium of the hand, a comparative relationship to a foot; and the particular parts of the foot [bear], through the medium of the foot, [a comparative relationship] to a hand—so that all [members] are comparatively related to the whole animal.) (Nicholas of Cusa 2001, 96, tr. Jasper Hopkins)

As mentioned in the introductory segment, Nicolaus de Cusa's thought can be related to Gotthard Günther's poly-contextural understanding of (cosmic) life itself. In fact, Günther, however, only referred to the earlier Cusanus—that is to say, to the latter's *early* rendering of the *coincidentia oppositorum* as an example for the "known Platonic pyramid of concepts" (Günther, n.d.). If he would have studied the later Cusanus, Günther surely would have realized that Cusanus had been a forerunner of his own ideas. Moreover, and as stated in the introduction, not only does Cusanus complete Günther's understanding in a way that Hegel's dialectic was unable to, but also the idea of a poly-contexturality of rather divergent and unique life-worlds already shines through Cusanus's thoughts here.

There is not enough space to elaborate on this in very much detail here. It should only be noted that—not only the humanity-related question of different world views in the context of different historical social systems and their philosophical traditions receives a surprising foundation here—even the (speculative) question of the nature and significance of possible extraterrestrial intelligences in comparison to the possibilities of human consciousness could be further contemplated by starting from Cusanus's thoughts. The key would be to see the infinity-manifesting, different subject–object worlds that Günther is postulating (1980, 87) in the sense of his speculative logic and the poly-contexturality of cosmic life in relation to what Cusanus reflects as (infinitely creative) *posse ipsum* (capability itself) in his latest phase. This term, as one might put it with Max Scheler (1874–1928), represents Cusanus's last and most evolved semantic *Zeigestab* (pointing rod) (Scheler 1921, 546) toward an ineffable (non-definable, infinite) creativity *which is at work in all possible subject–object universes* (as the former's representations) in the way Günther has postulated them (for the link to Cusanus, see also "*Preliminary Considerations*": Cusanus's term *microcosmos*—and Maturana's "multiversa").

Montaigne's View as an Implicit Forerunner of Poly-contexturality?

It should be mentioned that Michel de Montaigne (1533–1592) was inspired by Lucretius's ancient many-worlds theory. Stephen J. Dick points out that the years "1563–4 saw the publication in Paris of the Lambinus edition of Lucretius, a landmark in Lucretian studies because it remained the authoritative edition for almost 300 years" (Dick 1982, 46). Montaigne, who "was not an atomist or even a scientific thinker" (47), made a very special discovery though. He actually *broke through the veil of the mono-contextural fashion of seeing the world*: "Now if there are many worlds, as Democritus, Epicurus, and almost all philosophy has taught, how do we know whether the principles and rules of this one apply similarly to the others? Perchance they have a different appearance and different laws" (Montaigne 2003, 474; also quoted in Dick 1982, 47). This idea can be transferred and "upgraded" in the sense of our contemporary knowledge context and modern understanding. Every world, every star system (including binary or trinary systems) with its planets, and also every planet without a star—all of these actually form a unique consistent system of their own. When Montaigne speaks of "principles and rules," this is to be understood in the sense of 'particular characteristics' of a particular unique system, not in terms of general laws of nature in the modern sense. With Gotthard Günther, one could say that each solar system or vagabond planet represents a unique contexture (unique appearance, composition and sets or constellations of processes) in an infinitely complex poly-contexturality—or should we even say "omni-contextural" universe? The inspiration that we can derive from Montaigne's conjecture is that what we mean by 'life' might have to be understood differently in regard to each "individual" systemic contexture of the cosmos.

From Bruno, Galilei, Descartes, von Guericke, and Others to Newton and Kant

Nicolaus de Cusa's intuition that the stars are suns (possibly with other planets orbiting them) found its way to Giordano Bruno (1548–1600), whose efforts "concerning the infinite universe and innumerable worlds" (Bruno quoted in Crowe 2008, 39) "ended badly" (Silk 2006, 181). After Galileo Galilei (1564–1642) had concluded that "if the stars are suns and if suns are orbited by inhabited planets, then our universe must be extremely richly stocked with life" (Crowe 2008, 51), René Descartes (1596–1650), later Cartesian cosmologists (Crowe 2008, 65–67), and many others, such as Otto von Guericke (1602–1686), advertised the view that each star forms the center of a solar system and that extraterrestrial life must be infinitely distributed throughout the cosmos (Goerwitz 1985). Dick (1982, 106–141) analyzes the many-world theories of those influenced by Descartes and especially the views of Christian Huygens (1629–1695), Gerhard de Vries (1648–1705), and Bernard le Bovier de Fontenelle (1657–1757). Huygens published an "entire book devoted to the question of extraterrestrial life" (Crowe 2008, 98).

After that, "Isaac Newton (1643–1727) was perhaps the first to address the notion of gravity *in an infinite universe*" (Silk 2006, 182, italics DB). Moreover, those influenced by the Newtonian universe have further developed the idea of a plurality of solar systems (Dick 1982, 142–175). In a work that was published in 1722 and republished several times during the subsequent decades, the moral philosopher William Wollaston (1659–1724) writes of "the inhabitants of other planets" (Wollaston 1746, 143) in relation to the understanding that the fixed stars are

> rather so many *other suns* with their several regions and sets of planets about them [...] to beget in [those who are seeking a proper understanding] an apprehension of their unaccountable *numbers* and of those *immense spaces* that lie retired beyond our *utmost* reach and even imagination. (Wollaston 1746, 143, insertion in brackets DB)

In 1750, Thomas Wright (1711–1786), whose thoughts are also discussed by Crowe (2008, 130–138), then somehow reintroduced Nicolaus de Cusa's old idea in a new (Newtonian) "key," so to speak:

> Since as the Creation is, so is the Creator also magnified, we may conclude in Consequence of an Infinity, and an infinite all-active Power; that as the visible Creation is supposed to be full of siderial Systems and Planetary Worlds, so on, in like similar Manner [sic], the endless Immensity is an unlimited Plenum of Creations not unlike the known Universe. (Thomas Wright quoted in Dick 1982, 167, Fig. 12)

Wright must have been aware of Cusanus's *De docta ignorantia*, because he clearly referred to the latter's conviction that the "most sacred and supreme Divinity [...] [is] essentially present in all other Places as well as in this, as being a *Being* whose Center is every where [sic], and Circumference no where" (Wright 1750, 80, insertion in brackets DB). Some of Nicolaus de Cusa's basic ideas were available in English translation as early as the mid-seventeenth century: the above-stated idea is even expressed on the title page in Nicolaus de Cusa (1646)!

Mutatis mutandis, and in a very poetic fashion, Wright also reiterated Cusanus's idea of an interwovenness of all life throughout an infinite cosmos. At the same time, he emphasized its very different manifestations and forms in terms of planetary environments with divergent features, such as different atmospheres, etc. (Wright 1750, 81). This bears a certain systematic or thematic family resemblance to Montaigne's aforementioned remark. In the transferred sense, one could also say that possible extraterrestrial biospheres must be envisioned as distinct aspects of particular and unique planetary contextures (reflecting their own special conditions).

Regarding the possibility of intelligent alien life, Wright also quoted the aforementioned William Wollaston (without mentioning him by name):

> Besides, how can we tell but that there may be above us beings of greater powers, and more perfect intellects, and capable of mighty things, which yet may have *corporeal* vehicles as we have, but finer and invisible? Nay, who knows but that there may be even of these many *orders*, rising in dignity of nature and amplitude of power, one above another? It is no way below the *philosophy* of these times, which seems to delight in enlarging the capacities of matter, to assert the *possibility* of this. But, however, my own defects sufficiently convince me that I have no pretension to be one of the *first* rank, or that which is *next under* the All-perfect. (Wollaston 1746, 196; also quoted in Wright 1750, 82)

Under the background of the influence of Newton and others, Immanuel Kant (1824–1804) (see also the overview in Crowe [2008, 138–151]) introduces our present understanding of a—possibly—infinite multiplicity of galaxies:

> The size of a planetary world construction, in which the earth is hardly noticed as a grain of sand, sets the mind in astonishment; but imagine the delight and the astonishment caused when one is fathoming the immeasurable quantity of worlds and systems that fill the epitome of the Milky Way; and (imagine) how this astonishment increases as one is realizing that all of these immeasurable orders of stars constitute the unity of yet another number, the end of which we do not know, and which is perhaps just as incomprehensibly vast as the former—and which is, yet again, just a unit in a new correlation of numbers. We see the first links of a progressing relation of worlds and systems, and the first part of this infinite progression already shows what one should assume of the whole. There is no end here, but an abyss of true immensity. All capability of human concepts is submerged in this, even if it is raised by the help of number science. (Kant, n.d. [a], 256, tr. DB)

It has gone mostly unnoticed that Kant's general philosophical motivation seems to have been very much related to the assumption that humanity and all planetary life as such as well as the infinite universe *form one system* of cosmic life. Even Kant's famous question "*Was ist der Mensch?*" (Kant, n.d. [f], 25) (What is the human being?) has to be viewed from this angle.

In this regard, Kant's position is very interesting, because he seems torn between two positions, namely, between a more pessimistic and a more optimistic outlook. On the one hand, he sees the human being as a "crooked wood which cannot be made even" (Kant, n.d. [c], 100, tr. DB). Under the background of his assumption of a "countless multitude of worlds" (Kant, n.d. [b], 162, tr. DB) *as well as standing in awe before of the mystery of the life-bestowing force itself* (Kant, n.d. [b], 162), Kant therefore writes:

> If one asks whether the human species (which, if one thinks of it as a species of intellectual terrestrial beings in comparison *to those on other planets*, as a multitude of creatures [...] can also be called a race) [...] is to be regarded as a good or a bad race: I must confess that there is not much to brag about. (Kant, n.d. [e], 331, tr. DB)

This view is somewhat similar to the last quote by William Wollaston (the one also cited by Thomas Wright).

On the other hand, despite Kant's rather pessimistic view of our inability to evolve our interactions in proper moral and ethical reciprocity at least on a wider social or global scale, we cannot deny that Kant's philosophical works themselves *implicitly* express a sort of "principle of hope" (Bloch 1998), one might say, as a motivational aspect of his work. In my view, a spark of utopian optimism can be sensed in Kant's whole transcendental project itself. Despite the differences of content, time, and civilizational context, one should be able to recognize a basic "resonance" with the ancient Zoroastrian optimism (see the last main segment) that moral and ethical progress is possible.

According to Kant, the *cosmic* human being has the capability to transform its potential into a proper, ideal existence as *Vernunftwesen* (rational being) (Kant, n.d. [d], 444). It dawns on Kant that all human and non-human life must be an interconnected whole—on planet Earth but also beyond and with regard to the wider cosmic horizon. In his very last written thoughts in *Opus Postumum*, Kant therefore expresses his conviction that

> [t]he organizing force of our living and birthing globe has also organized the whole of the plants and animal species created for each other in a way so that they, the human being not excluded, act as links in a chain to form a circle. [...] *This means a hint at a world organization (for unknown purposes) even with regard to the system of the stars.* (Kant, n.d. [g], tr. and italics DB)

In learning to consciously partake in this cosmic context, human beings transform themselves into fully self-conscious rational beings. According to Kant, we are in command of the intellective processes of transcendental practical reason. These processes govern our possible experience. Thus, we have to assume that Kant's famous statement of an "ever new and increasing admiration and awe concerning the starry sky above me and the moral law within me" (Kant, n.d. [b], 161, tr. DB) is not just to be read as a rhetorical figure. As a *manifestation of cosmic life* in the particular sense of the above indented quote, that is, as a *cosmic "Vernunftwesen,"* we are in reach of a higher capability which expresses the self-organization of the world as such. The human being is able to realize a foundation of life which is "independent of the entire world of the senses" and which "goes into infinity," according to Kant (n.d. [b], 162, tr. DB).

In his 1804 version of the *Wissenschaftslehre* (*Science of Knowing*), Johann Gottlieb Fichte (1762–1814) stated that this understanding should lead to an absolute point of perfect convergence of theory and action—in borrowing from Nicolaus de Cusa one might say: to a *coincidentia oppositorum* of intellection and praxis—*in the unceasing inner certainty that* "there is only life, and this is eternal itself. *Whatever exists, is in life and is as eternal as life is*" (Fichte 2005, 77, tr. W.E. Wright, italics DB).

With these words, I would like to end the chain of examples of this particular chapter. We can see that also Kant and Fichte developed panenbiotic views. Further research in this direction will certainly provide more inspiration for the development of a *new trans-classical panenbiotic philosophy of cosmic life*. The further developmental history of this thematic field in German-speaking philosophical discourses of the 19th and 20th centuries would deserve another investigation, for example, with regard to the philosophies of Georg Wilhelm Friedrich Hegel (1770–1831), Friedrich Wilhelm Joseph Schelling (1775–1854), or the aforementioned Karl Christian Friedrich Krause. At this point, in the interest of space, further explorations in this direction of the history of philosoph*ies* have to be left pending.

Cosmic Life and Cosmic Light

The examples that have been presented in this article so far show that the thematic horizon of cosmic life is related to the topic of a constant self(re)creation and transformation of life as such. From a modern perspective, and in extending Kant's remark in *Opus postumum*, one might add that an "uninhibited" terrestrial biosphere is creating more and more complex systems (as its manifestations) and that this principle of progressive and adapting self-organization might even be extended "to the system of the stars" (Kant, n.d. [g], 570, tr. DB). In passing, I would like to mention that this understanding also means a contentual bridge to the modern cybernetic and systems theoretical approaches that have emerged since the second half of the 20th century (Rasch 1991, 66). In this direction, we already found that from the poly-contextural, trans-classical, and transcultural angle (see, e.g., "Poly-contexturality in the Theoretical Exploration of Cosmic Life"), we can think of life in a much more fluid way than in the sense of mutually exclusive, mono-contextural terms of 'inert' matter and 'living' matter. This has to be viewed in relation to our second, poly-contextural "*Globalperspektive*" (global perspective) (Luhmann 2008, 199), in the context of which we have seen that, since ancient times, the idea of an infinite cosmos is related to that of an infinite distribution of life. In the sense of such a conjecture, possible life "out there" is to be philosophically envisioned as a manifestation of boundless unity in an infinity of divergent forms and existences. Against this background, a third basic thematic horizon comes to the fore. It is triggered by the question: What could be the unifying factor of all life? This brings us to the third (transcultural) cornerstone of a possible future philosophy of cosmic life—namely, the philosophical topic of a correlation of cosmic life and cosmic light.

Historical Perspectives: Ancient Egyptian, Chinese, and Celtic Views

Since ancient times, light has always been a very important study object of philosophical reflection. At the beginning of the 20th century, the speed of light (in the sense of the modern concept of physics) became the general yardstick of modern cosmology. From the present perspective, and in a more general philosophical sense, the topic of 'light' deserves a special place in the discussion on cosmic life—namely, from a historical, intercultural, as well as contemporary systematic perspective.

It might be safe to say that in general, ancient peoples understood that the light of the sun, the moon, the visible planets, and also the light of the stars represented a (cosmic or encompassing) *conditio sine qua non* for their survival. The meaning of daylight has a positive connotation in ancient myths and also in the earliest written documents. Understandably, the diminishing of light during the night was seen as something negative. It meant the periodical emergence of a realm of real dangers, that is, wild animals or human enemies possibly approaching, etc. In addition, it provided the dark canvas for the projection of all kinds of mythological fears, and of a respective imagination of magic creatures lurking in the unknown, etc. At least, there were the moonlight and the light of the stars and planets.

The winter solstice-related festivals of peoples in colder climate zones were celebrated for the return of the more extended phases of daylight. In view of warmer climate zones, one can think of the sun religion of pharaoh Akhenaten (1427–1336 BCE). Here the light of the sun was described as the fountain of life. Likewise, in more orthodox ancient Egyptian lore, the god Ra stood for the rising sun and was glorified as "the giver of life." I would like to quote from a historical translation and interpretation, which is an important document in the history of Western Egyptology: "The benefit of life was linked to the rays of the sun, and is figuratively expressed on the monuments depicting the worship of Aten-Ra; the sun's rays there *emanate into hands that bestow the insignia of life on people*" (de Rougé und Brugsch 1850, 376, italics DB).

The ancient Chinese identified the earlier-mentioned principle of *yang* (see the subsection "Yijing (Book of Changes)") as a most basic "masculine" principle in the process of an unceasing living universe. It is also symbolized by the compass and the circle, which in turn were seen as symbolic representations of the sky. The ascribed function of this *yang*-principle was to "impregnate" everything on Earth with life. The earthly realm, in turn, was symbolized by the right angle and the square, which are emblems of *yin*. According to this view, the "impregnation" by *yang* still depends on the absorption and the nurture of the impulses by the "feminine" *yin* of the earth to fulfil its function. However, *yang* is identified as the life-*initiating* aspect and is associated with sunlight and daytime, while *yin* alludes to the comparative darkness of starry or moonlit nights.

Originally, the character *yang* 陽 was derived from the image of the southern side of a mountain and the northern bend of a river. These are the regions which receive the most light from the sun. The composite character (陽) contains the individual character for 'sun' and 'day' (日). The combinatoric symbolic matrix of the *Yijing* is said to have been created to "assist the '*luminous* gods' (*shenming* 神明)" (*Yijing*, n.d., "Shuogua," § 1, tr. DB). From the perspective of very ancient Chinese astrotheology, these can be identified as the time-bestowing and *light*-bestowing—and therefore *life*-bestowing—celestial bodies. The *Yijing* is conveying a code that describes how the *life-bestowing information of light in our planetary environment is constantly self-regulating and transforming itself in the form of human consciousness* (which is part of this life-bestowing light) as well as by way of all living processes as a whole.

To provide a third ancient example for the idea of life from light, we can direct our attention to ancient Celtic world views. Not unlike the ancient Chinese, the Celts envisioned their cosmos as consisting of a light-providing *and life-providing* realm and a complementary realm of darkness, "heaviness," and matter. *In between* these two cosmic dimensions of a light-bearing sky and light-absorbing darkness, (cosmic) *life flourishes as a third, intermediate realm.*

Antón Bousquet reconstructs this world view as follows: The highest dimension of the Celtic (living) universe is *albios* (ΑΛΒΙΟC). This means the luminosity or light that is found above darkness and gravity and even high above earthly life. *Albios* is the sky, in which light travels and which makes life possible at all. Furthermore, it is *albios* which enables *the living beings to contemplate their worlds.* The sky with its light is related to the idea of reflection and thinking. The other dimension of darkness was called *dubnos* (ΔΟΥΒΝΟC). The name of the third and intermediate dimension of cosmic life is *bitu* (ΒΙΤΟΥ). The word is etymologically related to the English verb '(to) be,' to its Russian equivalent быть, and to German *du bist* (you are). (English, Russian, German, and the ancient Celtic languages all belong to the Indo-European language family.) *Bitu* might best be translated as 'life and being,' *everlasting cosmic life as such. Bitu* exists on top of the great body of darkness (*dubnos*) (Bousquet 2018, 10). The integral *unity* of individual life and death at the root of never-ceasing universal *life as such* (see also "Cosmic Life as an Unceasing Transformative Self(re) creation") is symbolized by the god Biturix. This deity was envisioned as standing at the "threshold" of *albios* and *dubnos*. Biturix represented the overall balance and the equilibrium of all life and all living processes (Delamarre 2003, 76–77).

(Neo-)Platonic Traditions

For Platonic and Neo-Platonic thinkers (like, for example, Plotinus, who was mentioned in the second main segment already), the idea of a living universe and their conceptualizations of *light* were interlinked. For Plato, the sun meant an analogy or even a direct image of the highest idea of eternal and boundless goodness. As the rays and the warmth of the sun are the necessary condition for all life on the planetary surface (the 'telluric' realm), the spiritual light of the *idea* (ἰδέα) of the Good was seen as the origin of all processes of thinking, cognition, and experience.

Even the word ἰδέα (*idea*), which refers to the central principle of Plato's philosophy, is derived from the meaning of *seeing* and the appearance of something (visuality). It is directly related to 'light'. For Plato, the human eye is sun-like, and in the *Timaeus* he states: "The first of the organs [that the gods] constructed were the light-bearing eyes [...]" (Plato 2008, 35 [45b]). The eye enables us to see the stars and to grasp mathematical principles. It is a basic precondition for us to be able to finally discover philosophy and the highest principle of the good itself (Plato 2008, 38 [47a–b]). In view of these light-bestowing and light-bearing foundations, Plato in the *Phaedrus* defines life as self-movement(Plato 1913, 468–469 [245c–d). This definition is not limited to individual living beings. Besides the individual soul, which is self-moved, Plato in the *Timaeus* assumes that the whole cosmos is animated and self-moved (Plato 2008, 22 [34a]). He views the *whole universe as a perfect, immortal living being* (21 [32d–33a]).

These general points of view were developed further by Neo-Platonic thinkers since late antiquity. Based on Plato's views, Plotinus interprets light as the eminent life-bestowing representative of "the One–Good" (ἕν–ἀγαθόν) (Aubry 2020, 211), that is, *as the source of the living universe*.

> And the sun is also a paradigm of this, like the centre of the circle in relation to the light coming from it, and dependent on it. For the light is really everywhere with it and is not cut off from it. Even if you wanted to cut it off at one end, the light remains with the sun. (Plotinus 2018, 106 [*Enn*. 1.7(54).1])

It is important to mention the equivalence between an all-encompassing intellect and the world soul on the one hand and the luminous warmth-providing nature of sunlight:

> This cosmos, then, is plausibly said to be an image, always remaining dependent on its source [the One–Good]. The first [Intellect] and the second [(World) Soul] principles are at rest, and the third [nature] is itself at rest but is also accidentally in motion in the matter. For as long as Intellect and Soul exist, the expressed principles will flow out into this *form of* soul [the use of the lower-case 's' here is to indicate that this life principle of individual organic beings, which is identical with Plotinus's 'nature,' has to be discerned from the higher principle of the Soul of the cosmos as a whole]; *in the same way that, as long as the sun exists, there will be all the light that flows out from it.* (Plotinus 2018, 163 [*Enn*. 2.3(52).18], insertions in brackets and italics DB)

According to Plotinus, all life is participating in the process of cosmic consciousness in various degrees: Even plants have feelings, and the human ability of partaking higher levels of the world soul and to even actualize a (self-)reflection of the principle of the intellect (which means an unbound thinking of thinking or intellection of intellection as a reflection of the One–Good) is just a higher representation of a process of rational self-unfoldment which takes place in all forms of organic life (Bartosch 2023).

Another important Neo-Platonist, Proclus (412–485), then finally overcame Plotinus's dichotomy of a dark (absolutely light-devoid) and therefore evil physical matter (Bloch 1972, 149–150) and, on the other hand, the vision-providing light of the One–Good (Proclus 2003, 58, 65, etc.). Of course, light must also be present in living matter, and, at least for Proclus, matter cannot be designated as the source of evil (Opsomer and Steel 2003, 17). Everything is a representative *of* and everything is seen *in* the "light" of *the infinite principle of light*. This enabled John Scotus Eriugena (c.800–877) and Meister Eckhart (c.1260–1327/28) to envision nature and the living beings as material expressions and appearances of the highest and otherwise hidden principle of it all, the One–Go(o)d—and of its 'light' that is vivifying everything, that makes us see, and which represents an analogy of the spiritual "light" that is manifesting itself in the form of thinking.

Also for Nicolaus de Cusa, this *infinite* luminous principle of life is participated in self-reflection, that is, when we understand that *true (infinite) unity* can only be reflected as a boundless unity *through* difference or, put differently, as the *unity of unity and difference* (Bartosch 2015, 2022). As long as the life-bestowing principle, which is represented by those aspects of cosmic light (*mutatis mutandis* in the language of present-day physics: the electromagnetic spectrum) that we are planetarily attuned to (mostly in the visible spectrum and in the sense of electromagnetic radiation-based heat transfer), is thought of as being set apart from (finite) individual life and from our "personal" organic existence, it is mistakenly reconstructed in the form of a concept that also indicates a (countable) *unit* besides other unit*s*.

For Cusanus, this fundamental insight led to the formulation of his dialectical principle of 'knowing non-knowing' (*docta ignorantia*) (Mojsisch 2001; Bartosch 2015, 2021, 2022). In a transferred philosophical sense, Cusanus describes his first "aha-experience" (Karl Bühler, 1879–1963) in this regard as a "donation by the father of lights" (*dono a patre luminum*) (Nicolaus de Cusa 1932, 163; see also Bartosch 2015, 236). It is highly remarkable that Cusanus understands the whole (infinite) universe (populated with various forms of life) as a reciprocal act of *seeing*; it is "made of" creatures that experience finite consciousness in "their" particular perspectives and unfoldments of personalized time and space (in relation to their organic structures and processes), namely, as the respective *manifestation or appearance* of an infinite creator and an infinitely complex act of continuous creation. The act of seeing itself is understood as an intentional, processual act of meeting with the light of the One (*unum*), projected by the sun and other celestial bodies, that is, it is "viewed" as an integration with the infinite (visible light) under the conditions of finiteness.

More Systematic: Transcultural Foundations and Innovative (Unorthodox) Approaches

To move forward in view of the questions of what life is in general and how life on Earth might be related to the cosmos—which therefore might have to be considered as an infinite 'Life' (see also the chapter written by John B. Cobb, Jr.) as a whole (from the Gaia-hypothesis to the "Gaia–Uranus hypothesis," as one might also put it)—we urgently have to start transforming the relics of mono-contextural paradigms and belief systems that are hindering progress in so many fields. Seemingly unconnected disciplines and philosophical traditions have to be integrated into evolving poly-contextural metaverses of open and progressive thinking. Only in this manner will we be able to include the maximum amount of useful thoughts that ancient and present-day civilizations have to offer, establish new connections between them, and be inspired to explore an infinite universe from the angles of innovative observer pathways. This, of course, cannot mean a loss of critical differentiation, less to say arbitrariness in these processes of gaining new insights. The complexity of the system of understanding has to be raised on the basis of logic and established understandings of verification and falsification. At the same time, and as no theory can claim to represent an absolute truth (from this trans-classical perspective of poly-contexturality), mutual recognition or at least tolerance of divergent points of views and different hypotheses will mean an important foundation in the discourse on cosmic life.

Moreover, to follow the idea of poly-contexturality in science means to enable the development and unfoldment of a more *precise* method of thinking toward an infinite horizon of manifest possibilities, yet also a more pluralistic or "democratic" approach toward diverse propositions. It will allow us to deal with the enormous wealth of data and scientific interpretations unfolding right in front of our eyes in accelerating ways today. We will have to find a new way to "read" the insights with which artificial intelligence systems (AI) will provide us. Trans-classical science can also help us find a way to pose the right questions to make more effective use of the new possibilities. To develop a future philosophy of cosmic life, a sophisticated approach has to pave the way for a highly complex, pluralistic framework of different civilizational and cultural influences, in which all of these influences are connected without neglecting or dissolving their intrinsic differences. In such a fashion, it will be possible to bring to fruition their respective value for an overall understanding in view of the various general problem fields and thematic horizons of the topic of cosmic life.

Innovative theories in the field of the natural sciences and in transdisciplinary contexts (overcoming the long outdated reciprocal seclusion of fields in the natural sciences, social sciences, and the humanities) will only stand a chance in an inclusive and pluralistic environment. In view of a future philosophy of cosmic life, traditional and non-traditional systems of philosophy and sciences as well as their respective "truth-approximations" (within their particular contextural parameters) have to be cultivated *under the banner of mutual recognition* or, at least, tolerance—and at best: *mutual appreciation in diversity*. One philosophical task in this context will be to establish the logical foundations (also in the sense of examples of their practical application) to possibly relate all of the singular approaches of particular investigations and their 'truth-like' (Nicolaus de Cusa: *verisimile*) results to the more complex environments of their particular questions and problems in a fashion that enables further development and coordination of insights.

More Thoughts in View of Niklas Luhmann and Gotthard Günther

All the philosophical positions and approaches toward the problem horizon of cosmic life discussed here are manifestations of particular social systems of the past. They represent subsystems of philosophical communication in these encompassing processes. In view of Niklas Luhmann, one might say that the foundation of those (social) systems is represented by the acts of communication themselves, that is, in the sense of linguistic communication but also other forms of symbolic inter*action* (Bednarz 1989, x). All the historical views mentioned here represent products of the philosophical subsystems of particular societies' engagements with a natural, respectively, cosmic environment which is to be assumed as being far more complex than the reflecting discourses and their thought patterns themselves. From a *philosophical* point of view, we cannot say that one of these ancient or more recent views is valid to a higher degree than another, because, as methodologically supported efforts, these contextures are all similar in that they all approximate the realm of an infinite (cosmic) creativity, that is, a universal noumenon which is the basis of an inexhaustible scientific creativity and which can and will never be brought to a final (finite) conclusion. From our perspective of knowing non-knowing (*docta ignoratia*), that is, to know that to "know" the infinite (by way of finite concepts of reasoning) is impossible (Bartosch 2022), we are enabled to switch from one of the historical communicative memories to that of other socio-historical philosophical subsystems, namely, as integrated observers in a trans-classical generativity of understanding. Of course, this does not mean to ignore scientific progress, but also the understanding that some progress can lead to a dead end. If we nurture all the divergent contextures at once, we will have more foundations and materials to alter or refreshen our paths of understanding in most progressive ways.

To survive and to achieve ecological equilibrium in an expanding global system that is approaching its hypercomplex cosmic environment in more and more (and often ecologically devastating) ways, we have no other option than to make our present and more interconnected subsystem of philosophical and scientific communication *more complex* (on complexity in general, see also Woermann 2016). From the perspective of the philosophy of cosmic life, this problem cannot be confined to the 'telluric' environment (see "Preliminary Considerations") of organic life on Earth. It goes without saying that this complexity increase is to be imagined in the transferred sense of an "extropic process." We are talking about an "organized" (Rasch 1991, 65)—or better: about a *self-organizing* increase in complexity.

> In *Ecological Communication* Luhmann uses the concept of 'resonance' to designate this system/environment interplay. But while complexity—environmental complexity—is always a problem for a system, at the same time it is the key to its solution through the increase of the system's own complexity" (Bednarz 1989, ix).

The most important question in this regard is the one concerning cosmic life. The complexity that the new problem horizon provides us with will help us to raise the complexity of our communication in an interconnected planetary social system. A first step is to bring together the views of different historical manifestations of divergent (local) traditions of philosophy. In this context, it is important to apply a logic which is *not* leading toward an absolute, or the "One Truth," or *the* (allegedly) one and only exclusive scientific paradigm. We need to refer to an approach that can help us constantly interlink our different perspectives and viewpoints in the sense of a vast, constantly evolving, and self-unfolding field of transdisciplinary and transcultural communicative acts. Again, the "connecting tissue" is provided by the application of the implicate-logical foundation of the unity of unity and difference (Bartosch 2022) in view of the problem horizon of cosmic life itself.

The metaphor of the horizon, which is to be found in the philosophy of Nicolaus de Cusa already, is also to suggest that there is *always* something new to discover, we just have to constantly move toward it, namely, to strive to get past the respective present horizon. Philosophy and science are not to be misinterpreted as forms of religion; they should never be misunderstood as fixed belief systems. Thus, philosophy and science are not finalizable (Bartosch 2019). Both have to evolve and to "shapeshift" from contexture to contexture in unique observer pathways (also Maturana 2003, 57–58) of exploration. These transformations imply transdisciplinary connections, related segments of pathways of interlinked understanding. Philosophy and science are not based on the habitual belief in an unfounded narrative. Dogmatic settings are not within the scope of true science and philosophy. The problem of cosmic life, if handled properly, provides the key to enter this "space" of true, that is, *ever-evolving* understanding.

I have referred to Gotthard Günther's approach of poly-contextural logic, which can—and here we have to move with Günther past Günther—itself be viewed as a manifestation of an infinite principle which is "worded" (Cusanus: *verbatur*) in each and every word *in an ineffable way* (Nicolaus de Cusa 1959, 54). Instead of trying to approach an absolute by means of higher and more comprehensive synthesis, for example, in the sense of an alleged "theory of everything," the *true* absolute of philosophy is *already* implicitly "speaking" in the unfoldment of the following approach: "Thus it becomes clear that it is not the task of the mediation to reconcile the simple value opposition of positivity and negativity at all in progressive stages, *but to expand the relational content and the interpretative capacity of reality*" (Günther 2005, 7). There is no single absolute point of truth, just a never-ending expansion (or infinite expanse) of truths in a possibly infinite number of reflections and a permanently shifting and expanding horizon of finite and particular fields of knowledge.

In essence, the whole unfolding of individual observer pathways through the conundrums of a poly-contextural universe is driven by the paradoxical statement that "there is no absolute truth." *The statement itself implies what its meaning denies.* No contexture, no single view, can convey the absolute truth of that which runs through all particular "truths" in analogy to the indefinite *dao* 道 or *apeiron* (ἄπειρον) constituting all singular number-units (which form an infinite set of integers) as their immanent *and* yet unfathomable (transcendent) foundation. But the absolute truth runs through the difference between all contextures in a negative sense. It cannot be known in the sense of particular knowledge, but it can be known that it cannot be known. In this sense, the statement that there is no absolute truth represents an absolute truth *ex negativo*.

It is under this infinite background of self-reflexive (self-referential) negativity that our open and yet self-contained system of philosophical and scientific communication is supposed to be raised in terms of its complexity. We have to explore our problem horizon of cosmic life in as many directions as possible, namely, to approach and to obtain "further communicative alternatives" (Bednarz 1989, x). I would like to repeat myself deliberately here: An extended Güntherian approach (also in the sense of what has been stated earlier in the introduction) can provide us with the perfect (implicate) logical (Bartosch 2022) framework to relate a vast field of historical and present-day approaches to each other from a transcultural and transdisciplinary angle.

We have seen that there are certain basic "commensurable" topics of 'cosmic life' which, taken as comparable systematic content from our meta-perspective, lead to the insight of permeable problem horizons of different systems of communication (ancient Indian, Chinese, Persian, Greek, etc. philosophies). In other words, the transcultural survey has shown that there is a certain kind of "resonance" with regard to the *problem* of cosmic life amongst the philosophies of many ages and places. It became clear that the idea of an ever-living nature of life as a whole (which is implying individual death and also consummation by other individual forms of life), ideas of a spacially infinite proliferation of life or, at least, many-life-world theories, and, last but not least, the light-derived nature of life have appeared even independently in several ancient but also later philosophical contexts. This provides a foundation to establish more holistic and pluralistic frameworks and poly-contextural networks, which then would allow for the inclusion of "unorthodox" *modern theories*, that is, points of views which have been neglected or underrated, because they did not fit the respective mono-contextural "standards." Upon closer inspection, such "standards," that is, paradigmatic or dogmatic scientific habits and social structures of belief, are based on the assumption that all knowledge might one day converge on a single point of truth, mostly in relation to factors of possible benefit that lie outside of pure scientific knowledge interests themselves.

Photons, Life, and Consciousness

Against this background, ancient views on the correlation of natural light and life and consciousness might be seen as indirect pre-scientific inspirations or forerunners of advanced, yet marginalized scientific theories of our times. Not in the least in the sense of a further general conclusion, I would like to postulate this additional view here:

Life, respectively, cosmic life is to be viewed as an aspect of light. In the light, there is life. Light must be more complex than life. A basic principle of life (as part of light) is its self-induced motion or movement. If something is radiating light from itself, it is part of (cosmic) life. If something only reflects or absorbs light, it is not life, respectively, it is not alive. If something actively radiates visible or non-visible forms of light (electromagnetic radiation, photonic energy), *it participates in cosmic life. Human consciousness means such a radiation of energy, of 'light' in the extended sense.*

The first two sentences can also be related to the common understanding that.

> [i]f you stir up simple non-organic molecules like water, ammonia, methane, carbon dioxide and hydrogen cyanide with almost any form of intense energy, ultraviolet light for instance, some of the molecules reassemble themselves into amino acids, a result demonstrated about thirty years ago by Stanley Miller and Harold Urey. (Hoyle 1983, 18)

However, philosophically speaking, we cannot infer from this

> that the correct arrangements of amino acids, like the orderings in enzymes, can be produced by this method. [...] Nevertheless, many scientists have made this leap [...] in spite of the obviously huge odds against such an event having ever taken place on the Earth, and this quite unjustified conclusion has stuck. (Hoyle 1983, 18)

In fact, Hoyle's criticism is still right, because this leap represents a logical fallacy: We cannot conclude from the aforementioned physical fact, where photonic energy is acting on inert matter, on the *active* processes of *living* autopoiesis and self-induced movement. This would present a category mistake. This, by the way, holds also true for more recent findings, for example, according to which "[a]queous microdroplets enable abiotic synthesis and chain extension of unique peptide isomers from free amino acids" (Holden et al. 2022). The fallacy becomes even more striking, if we take into consideration that—and this is a proven fact—the living human body "directly and rhythmically emits light [and that the] diurnal changes in photon emission might be linked to changes in energy metabolism" (Kobayashi et al. 2009, [1]).

Regarding my postulate at the beginning of this subsection, we have to search for theories in the natural sciences which explore the general philosophical idea that life is a phenomenon participating in 'light' (in the most general sense of a comparative category) in experimental contexts and under the background of modern scientific world views.

The challenge in this context is *that we have to broaden the concept of life itself* —namely, to alter away from those "encapsulations" of thinking that can be traced back to an era when the world was explained as a mechanic device (e.g., Julien Offray de La Mettrie [1709–1751]) . Generally speaking, the favorable alternative is that the universe itself is '*life-centered*' (Attila Grandpierre). 'Life' is also part of that which is not in the state of "solid" matter and which cannot be explained mechanistically. Moreover, 'life' is primarily also partaking in radiation processes and the derived forms of *energy transformation*.

It is a most remarkable, yet widely ignored fact, that, "like supernovas, the human body emits light before dying" (Isabel 2021). Here, we can therefore even refer back to Moses Hess's parallelization of the death of stars (supernovae) and the death of Earth-based biological organisms that was introduced at the very beginning. The outburst of light at the moment of death is so immense that it is held responsible for particular near-death experiences.

> If this excess of bioluminescent photon emission exceeds a threshold in retinotopic visual areas, this can appear as (phosphene) lights because the brain interprets these intrinsic retinotopic bioluminescent photons as if they originated from the external physical world. (Bókkon et al. 2013)

Before we continue, I hasten to say that, at least on this occasion, the task cannot be to speculatively probe the ground for alterative theories of the origin of life, for example, how light/radiation and the emergence of organic matter in combination with water—for instance, also in the context of asteroids like Itokawa (Chan et al. 2021)—might help us to shift the attention away from the common, if I may say so, "telluric fixation", or in what sense abiogenesis could be replaced by a completely different hypothesis. Philosophically speaking, before one can think of tackling such kinds of more advanced questions, we have to discuss more in depth what 'life' means, and we will have to develop and unfold an appropriate philosophical concept of 'life,' namely as an entrance point to evolve our further understanding in various ways from there.

In addition, we have to deepen our understanding of 'light' in a more extended philosophical fashion. Therefore, from the "multi-angle" of the poly-contextural approach, I would like to direct the attention to the fact that there are also unorthodox theories in view of the correlation of light and life, which are important to know. These have mostly been ignored so far —not because they lack the possibility of validation or because they were falsified, but for the simple "cultural" reason that they do not blend in with the "one-size-fits-all mentality" of paradigm-bound habitual belief structures in the social subsystems of the particularized and very often "industrialized" sciences of our day.

However, these innovative approaches must not be ignored. In our present context, and especially from the philosophical angle, such kind of perspectives can help us to evolve the concept of 'cosmic life' further, that is, the concept that we will also need to develop the question of possible origins at a subsequent stage. To enable progress, we have to leave the comfort zone of temporary paradigms and "scientific belief" (Bartosch 2019). In analogy to learning a new language, scientists must be "allowed" to think in non-habitual ways and even to make mistakes or to risk following wrong tracks without being stigmatized or outcast from the scientific community. If scientists are not empowered to think outside of the box and end up as mere "publication workers" to raise the level of university rankings in relation to mere quantities of "output," then the original meaning of science is weakened to a great extent. To make balanced progress toward a better understanding of (cosmic) life, we have to observe and to recognize heterodox and novel approaches as well. They *even should be cherished for what they represent*: the possibility of progress *that is the core characteristic of all scientific endeavors in the appropriate sense of the word.*

Biophotons: Fritz-Albert Popp

Regarding the problem of *cosmic life*, the topic of biophotons seems highly promising and should be investigated again (and maybe with new methods and advanced experimental systems at hand). According to the biophysicist Fritz-Albert Popp (1938–2018), there are two basic theories regarding biophotons: Both the physicalist biochemical theory (BCT) and the more innovative coherence theory (CT), which was developed by Popp himself, accept that biophoton emission is a general phenomenon of living systems. The physicalist biochemical theory claims that the phenomenon is just some kind of "'waste' of the metabolic events taking place permanently within the cells," while "the CT points to the low intensity as an indication of non-classical light which may display even sub-Poissonian photocount statistics and may thus provide an optimized optical communication channel in biological systems within living matter of 'optimized' high optical density" (Popp, n.d., 2–3; see also Bischof 1998 and Creath and Schwartz 2005). Also from this angle, (cosmic) life must *be much more than just the sum of the single molecules that constitute a living system* (Maturana 2003, 105).

That this theory was ignored by a majority of today's (physicalist) contemporaries does not mean that it has been falsified, to say the least. A theory is not falsified because it might conflict with common views, or because it exceeds the imagination or narrow set of accepted possibilities of a majority of "textbook-believers." It is no secret that possible lack of acceptance of innovative or even "heterodox" or "counter-paradigmatic" approaches can be related to social and communicative aspects of science communities. This has to be taken into account here, too. One might think of the rejection of Galilei's proposals in the socio-historical context of his day. More advanced scientific theories (and sometimes even accidental findings or certain inventions in relation to that) might also threaten a respective contemporaneous economic, political, or religious powerbase, and it is often the case that they are ignored or suppressed or "neutralized" for this simple "human" reason. We also have to keep in mind that the scientific community is posed with the question of the allocation of research funding in these contexts (including the related "allocation battles" in these scientific communities themselves).

However, nothing hinders us to pursue the idea in the present context. Please recall that according to the above-stated definition, everything that is emitting light by itself has to be counted as partaking in cosmic life in a much more comprehensive and even trans-classical sense. A proposal from the philosophical side would be to pursue the biophoton theory against the background of its wider cosmo(bio)logical implications. In this sense, biological theories of self-consciousness, such as those in the context of the concept of autopoiesis by Humberto R. Maturana and Francisco Varela (e.g., Maturana 2003), would have to be further developed today—especially with regard to a possible non-chemical form of communication in living bodies (as proposed and documented by Popp and others). In so doing, a new trans-classical science of (cosmic) life could also be reconnected to philosophy as well as to its many histories (and even mythological pre-histories) in the most fundamental sense of a new form of global philosophy which is embedded in a plurality of unfolding transcultural and interdisciplinary paths of thinking. The ancient Platonic understanding that consciousness, the mind, is some kind of invisible light is gaining a whole new scientific dimension from here—especially also if we keep in mind the additional aspect that the ancient Chinese code of the *Yijing*, related to the basic idea of life-bestowing interplay of light (*yang*) and darkness/absorption of light (*yin*), presents us with a direct parallel to the code-structure of the DNA (see "Yijing (Book of Changes)").

If one pursues the direction of Popp from a more philosophical angle, the coherent emission of biophotons from living cells can further be reflected as the sources of consciousness. By "[c]ombining the ultraweak biophoton imaging system (UBIS) with the biophoton spectral analysis device (BSAD)," it has been shown

> that glutamate-induced biophotonic activities and transmission in the brain, which has recently been demonstrated as a novel neural signal communication mechanism, present a spectral redshift from animals (in order of bullfrog, mouse, chicken, pig, and monkey) to humans, even up to a near-infrared wavelength (~865 nm) in the human brain. This brain property may be a key biophysical basis for explaining high intelligence in humans because biophoton spectral redshift could be a more economical and effective measure of biophotonic signal communications and information processing in the human brain. (Wang et al. 2016, 8753)

In the sense of further "food for future thought," it should also be remembered that Max Planck (1858–1947) said that "[…] photons which constitute a ray of light behave like intelligent human beings: Out of all possible curves they always select the one which will take them most quickly to their goal" (Planck 1950, 178). And, furthermore, photons have been proposed as carriers of consciousness in general (Romijn 2002; Moro et al. 2022). From our angle of a poly-contextural interpretation of existence and of our (self-)conscious reflection patterns (as system observers), it might also be noted that the wave–particle dualism of light, respectively, electromagnetic radiation represents a most basic poly-contextural phenomenon itself. It would also be worthwhile to transfer or, alternatively speaking, to project Niklas Luhmann's concept of *strukturelle Kopplung* (structural coupling) into this much broader as well as physics- and biology-related problem horizon. In my view, there is a lot to be derived from these angles in regard to a future panenbiotic concept of cosmic life.

Connecting Dots in the Light of Solar Astronomy: Attila Grandpierre

In the sense of his concept of comprehensive science, the Hungarian solar astronomer Attila Grandpierre shows one way how the concept of life on Earth can be extended toward a scientific vision of cosmic life. Grandpierre's starting point, which somehow extends the idea of a universal intentionality for life (see the second Schrödinger quote in the subsection on Uddalaka Aruni), is as follows:

> We found the analogy of solar activity in biological actions like bending our finger. Indeed, when we bend our finger, we act from the global level of our mind to the local level of our finger. The causal chain of solar activity starts from beyond the gigantic network of physical causes, extending to the entire observable universe. The Sun initiates biological causes from a deeper layer of the Universe existing beyond the quantum vacuum by creating virtual particles suitable for realizing biological causes, namely, regenerating solar activity. Both solar activity and bending our finger are genuine self-initiated, self-governed actions involving top-down causation. In this way, we have developed a complete theory of genuine solar activity that is called the Helios Theory [...]" (Grandpierre 2018, 121).

From the systems theoretical angle, one might alternate this understanding in the following fashion: The sun is showing one of the core characteristics of all life—*autopoietic behavior*. Somehow, in a new and timely way, we are, *mutatis mutandis*, in touch with Moses Hess's intuitions from the very beginning of the present chapter.

What is even more fascinating in this context is Grandpierre's rediscovery and further development of the biological ideas of Ervin Bauer (1890–1938) and the application of the latter's systematic biological definition of life (e.g., Bauer 1935) to the sun.

To explain the still rather unknown or forgotten ideas of Bauer, one can refer to the physicist Erwin Schrödinger at first, namely, as a contentual preparatory stage for our understanding—as a "mediator," so to speak. In his famous work *What is Life?* (1944), which inspired the discovery of the DNA-molecule, Schrödinger held the view that "[...] living matter, while not eluding the 'laws of physics' as established up to date, is likely to involve 'other laws of physics' hitherto unknown, which however, once they have been revealed, will form just as integral a part of science as the former" (Schrödinger 2013, 68). Schrödinger also discussed the "characteristic feature of life" (69), which he, to stress this important point again, describes as the avoidance of "the decay into thermodynamical equilibrium (death)" (73) by "feed[ing] on 'negative entropy'" (70; see also the Schrödinger quote in the subsection on Uddalaka Aruni).

It seems that Schrödinger was not aware of the works of Bauer. Bauer had published in German (1920) (which would have been accessible to Schrödinger) and in Russian (1935). It also counts as one of the greatest tragedies in the younger history of science that Bauer was killed during the Great Terror in the Soviet Union in January 1938. He had not published in English, and his important ideas were forbidden in the Soviet Union in 1937. His posterior rehabilitation by Soviet authorities in 1954 did not help prevent his theory from being forgotten until rather recently. However, it shall never be forgotten that 24 years before Schrödinger's *What is Life?*, Bauer published the following, utterly important insight:

> Every living being is characterized by the fact that it is a system which is not in a state of equilibrium in the given environment and is arranged in such a manner that the energy

sources and forms of energy of its environment are transformed in the same into such forms of energy which, in the given environment, act against the occurrence of the state of equilibrium. (Bauer 1920, 10, tr. DB)

On the basis of Attila Grandpierre's and other Hungarian scholars' works, Bauer's ideas and research have recently been reintroduced to the philosophical and scientific discussion. Other than Schrödinger, who merely saw the sun as a distributor of free energy for the biosphere, Grandpierre has been able to extend the concept of life to the sun itself—which represents an idea that might very well turn out to be one of the most important insights of our time:

[If] the essence of life can be formulated as the following: the Bauer principle (BP) is manifest in virtual interactions which generate biological couplings leading to investment of work ΔW that generates thermodynamically uphill processes increasing extropy ($\Delta \Pi > 0$) [...] [and by a]pplying the Bauer principle for the Sun, we found that the Sun is a living organism. We are led to recognize a cosmic life form in stellar activity cycles. Then we generalized the Bauer principle and found new kinds of cosmic life forms like the microscopic, intermittent and hidden life forms. We found that the first principle of biology is able to be manifest in the whole universe through virtual interactions. This result led us to recognize a new *cosmic life form present in the vacuum that we call universal life.*" (Grandpierre 2008, 369, italics DB)

From here, again, (cosmic) life can be seen as an element or as a manifestation of cosmic light—or put differently: of electromagnetic energy (Sanders 2014). Grandpierre proposes that as it is widely accepted that living beings can be defined as natural systems governed by information and "since solar activity is governed by its information content corresponding to its lawful algorithmic complexity, the Sun is a living system" (Grandpierre 2008, 381). Farhadi et al. (2007) show "that human cells in culture could synchronize their internal chemical processes even though they were mechanically, chemically, and electrically isolated from one another. The cells, it seemed, were communicating through the exchange of photons" ("The Emerging Field of Biophotonic Communication," 2009). As observed in the case of fish eggs and frog eggs, forms of organic life have already been shown to communicate via "photon radiation in form of short quasi-periodic bursts [...] [by means of a] communication mechanism [...] similar to the exchange of binary encoded data in the computer nets via the noisy channels" (Mayburov 2012). "Certainly, [...] biological life on Earth has already figured out how to utilize the astronomically huge flow of biologically useful information reaching the Earth from the Sun" (Grandpierre 2008, 381).

In my view, Grandpierre's use of the word 'flow' is important here. If we want to understand how (cosmic) life is also a characteristic of 'light' (in the sense that has been explored so far), we should not ask so much what light is *but "why" there is light.* We should not focus too much on the physical properties of light. Instead, we also have to see the *tendency* or the *tendencies* of light. In the transferred sense, we could ask: Which "role(s)" does it tend "to play"? A river cannot not just be defined by saying that it is made out of H_2O but by its particular characteristic flowing tendency *as a whole*, various characteristics under certain "circumstances," the life it nurtures, the substances it carries, its environments, etc.—and its embeddedness in the water cycle, etc. *Likewise, we have to understand light in the sense of an energy transfer that starts somewhere and reaches somewhere.* Against the background of Grandpierre's remarkable solar astronomical findings (in view of Bauer's concept of life) and also in view of Schrödinger's old assumption that the sun provides the free energy for the complexity increase that is characterizing the life of the planetary biosphere as such (Schrödinger 2013, 74; Rasch 1991, 66–67), it seems promising to me to think about 'cosmic life' as being part of this 'flow' of cosmic light—not just on a passively receiving side, but also as a (self-)conscious, active transmitter of photonic energy (and information).

Closing Remarks

As an open, ever-evolving, and at the same time self-referential system of thinking, *a future philosophy of cosmic life will necessarily always have to be in the making*. In the sense of a perpetually self-organizing complexity, this poly-contextural discourse will have to bring together as many understandings of 'life' from as many philosophical or spiritual heritages as one can with as many innovative developments in present-day philosophy and transdisciplinary (trans-classical) approaches of the sciences as possible. Very often progress in various directions is inhibited due to explicit or even unwanted dogmatism, namely, when certain habitual convictions get in the way. Even scientists and philosophers can be trapped by biases which might unknowingly derive from non-scientific and non-philosophical narratives (certain "atavisms" of outdated points of view or ways of thinking or believing, as one might also put it). These often blur the vision for novelty, and thus for what is deemed thinkable, imaginable, or possible. The understanding of what represents "proper science" or philosophy is often implicitly misaligned due to influences which cannot be identified as either scientific or philosophical themselves.

Of course, to vouch for novelty does not mean to diminish or even neglect the achievements of well-accepted and existing foundations of the natural sciences or philosophical disciplines. *On the contrary*, without the *existing* foundations and theories, progress would not be possible. But a second condition for progress is that these foundations are open for changes or extensions and further inclusions to increase the systemic complexity of our communication and our particular reflections and understandings.

I have already emphasized that 'trans-classical science' does not mean an antithesis or an exclusion but, on the contrary, the *extension* and inclusion of 'classical science' in the new form of thinking. Metaphorically speaking, the trans-classical means the "Cambrian Explosion" from the mono-contextural to the poly-contextural; it means an integration and "uplifting" of all the existing methods, ways, and understandings toward a much more comprehensive field of transdisciplinary and transcultural world-philosophical communication—without neglecting the importance of the prior scientific methods and achievements so far. The poly-contextural foundation is not less *but more scientific*. One can also put it as follows: The trans-classical approach is driven by the insight that the statement that "there is no absolute truth" represents a most basic aporia that cannot be overcome. It is a paradox, because it claims what it abnegates, that is, *by its refutation*. The only way "to measure up" to this most basic meta-paradoxical *principle* of understanding is to constantly *unfold* it as best and as thoroughly as possible. All knowledge must be valid knowledge, but it can never be claimed to be absolute knowledge or absolutely complete knowledge. Therefore, to say this again, it will always have to be *in the making*.

If we truly want to tackle the question of cosmic life, an exclusive decision for one theory will not be sufficient. We have to cultivate an open and expanding poly-contexturality of interconnected views of *different* perspectives and alternatives. Like some of the theories and points of view that have been discussed here, these might not always be commensurable to, for example, current (mono-contextural structures of belief in) one-world theories of life or advanced visions of abiogenesis. But as long as they are consistent in themselves and open for verification, falsification, or further development, they should not be marginalized, suppressed, or even excluded from the general discussions. And, perhaps, sometimes different perspectives need to be made complementary in order to reach new levels of understanding. Remember that the logical *connection* of all kinds of divergent (orthodox and heterodox) theories (and their variations maybe) as well as even radically new points of view can be found in the hidden foundation of all thinking itself, that is, in the "implicate logic" (Bartosch 2022) of the unity of unity and difference itself (see "Extending Classical Science Toward Transcultural and Trans-classical Views"). It is this very foundation that enables us to distinguish between true and false in the first place. The activity and ability of theory-building on that basis is unrestricted. It represents the openness which is conveyed by one of the most important and oldest assumptions of philosophy in ancient China (Confucius), Greece (Socrates), and probably in other ancient cultures of thinking: "I know that I do not know (yet)" (see also Bartosch 2021). Current status quos in various branches of the sciences or philosophical discourses are not a private property that can be kept from changing or from being developed further in the minds of others. The more different yet consistent reflections we can apply, the better. Only in this way can we approach the immensity of the cosmos and of life. Especially with regard to the burning questions of the future, we have to widen, to extend, and to pluralize the scope of our perspectives and understandings in most extreme (and courageous) ways. Mathematically speaking, there is not even just one infinity. To diminish the pluralism of possible scientific views in taking note of the unknown of a particular problem by any implicit or explicit (mono-contextural) form of dogmatism always resembles the attempt to explore an island without being able to move around. To unfold a growing poly-contextural understanding of *cosmic life*, one has to be open to exploring not just one "island," one scientific subdiscipline, current theory, or scientific point of view, but many "lands and seas."

References

F. Aaserud, A complementary relationship: Niels Bohr and China. Niels Bohr Archive (n.d.), https://www.nbarchive.dk/doc/Aaserud-confucius.pdf. Accessed 18 Oct 2022

L. Anagarika Govinda, Introduction, in *Schönberger* (1992), pp. 15–23

Aristotle, *On the Heavens*. Greek-English edition, with an English translation by W.K.C. Guthrie. The Loeb Classical Library, vol. 338 (W. Heinemann, London; Harvard University Press, Cambridge, MA, 1939)

Aristotle, *On Sophistical Refutations. On Coming-To-Be and Passing-Away*. Greek-English edition, with English translations of both texts by E.S. Forster. *On the Cosmos*. Greek-English edition, with an English translation by D.J. Furley. Loeb Classical Library, vol. 400 (W. Heinemann, London; Harvard University Press, Cambridge, MA, 1955)

Aristotle, *The Metaphysics, Vol. I-IX*. Greek-English edition, with an English translation by H. Rackham. The Loeb Classical Library, vol. 271 (W. Heinemann, London; Harvard University Press, Cambridge, MA, 1956)

G. Aubry, An alternative to Cartesianism? Plotinus's self and its posterity in Ralph Cudworth, in *Self-Knowledge in Ancient Philosophy: The Eighth Keeling Colloquium in Ancient Philosophy*, ed. by F. Leigh (Oxford University Press, Oxford, 2020), pp. 210–229

D. Bartosch, *»Wissendes Nichtwissen« oder »gutes Wissen«? Zum philosophischen Denken von Nicolaus Cusanus und Wáng Yángmíng* (W. Fink, Paderborn, 2015)

D. Bartosch, Karl Jaspers' *philosophischer Glaube* (philosophical belief) and Wang Yangming's *zixin* 自信 (self-believing). Minima Sinica **31**, 45–64 (2019)

D. Bartosch, The understanding of understanding: a philosophical reflection from a transcultural perspective. Int. Commun. Chin. Cult. **8**(1), 121–143 (2021). https://doi.org/10.1007/s40636-021-00211-7

D. Bartosch, Transcultural philosophy and its foundations in implicate logic. Asian Stud. **10**(3), 107–126 (2022). https://doi.org/10.4312/as.2022.10.3.107-126

D. Bartosch, Plotinus and Wang Yangming on the structures of consciousness and reality: a transversal prospection in view of affinities of their positions. Asian Stud. **11**(1), 91–135 (2023). https://doi.org/10.4312/as.2023.11.1.91-135

E. Bauer, *Die Grundprinzipien der rein naturwissenschaftlichen Biologie und ihre Anwendungen in der Physiologie und Pathologie*. Vorträge und Aufsätze über Entwicklungsmechanik der Organismen (J. Springer, Berlin, 1920)

E. Bauer, Э. С. Бауер, *Теоретическая Биология* [*Theoretical Biology*] (Издательство Всесоюзного Института Экспериментальной Медицины, Москва, 1935)

J. Bednarz Jr., Translator's introduction, in *Luhmann* (1989), pp. vii–xvi

M. Bischof, *Biophotonen: Das Licht in unseren Zellen*, 9th edn. (Zweitausendundeins, Frankfurt am Main, 1998)

E. Bloch, *Das Materialismusproblem, seine Geschichte und Substanz*. Werke, Band 7 (Suhrkamp, Frankfurt am Main, 1972)

E. Bloch, *Das Prinzip Hoffnung*, 3 Bände, 5. Auflage. Werke, Band 5 (Suhrkamp, Frankfurt am Main, 1998)

I. Bókkon, B.N. Mallick, J.A. Tuszynski, Near death experiences: a multidisciplinary hypothesis. Front. Hum. Neurosci. **7**, article 533 (2013). https://doi.org/10.3389/fnhum.2013.00533

A. Bousquet, *The World of the Gauls: Foundation(s) of a Celtic Philosophy* (Koadig, 2018)

H.G. Bronn, *Handbuch einer Geschichte der Natur*, Band 1: *Einleitung. I. Theil: Kosmisches Leben. II. Theil: Tellurisches Leben*. Naturgeschichte der drei Reiche, Band 13 (E. Schweitzerbart, Stuttgart, 1841)

H.G. Bronn, *Handbuch einer Geschichte der Natur*, Band 2: *III. Theil: Organisches Leben*. Naturgeschichte der drei Reiche, Band 14 (E. Schweitzerbart, Stuttgart, 1843)

H.G. Bronn, *Handbuch einer Geschichte der Natur*, Band 3: *III. Theil: Organisches Leben (Schluß). Index palaeontologicus oder Übersicht der bis jetzt bekannten fossilen Organismen*, bearbeitet unter Mitwirkung von H.R. Göppert und H. v. Meyer. *B. Enumerator palaeontologicus: Systematische Zusammenstellung und geologische Entwickelungs-Gesetze der organischen Reiche. IV. Theil: Vernunftleben.* Naturgeschichte der drei Reiche, Band 15 (E. Schweitzerbart, Stuttgart, 1849)

M. Boyce, *A History of Zoroastrianism*, vol. 1: *The Early Period,* 3rd edn., with corrections (Brill, Leiden, 1996)

Q.H.S. Chan, A. Stephant, I.A. Franchi, X. Zhao, R. Brunetto, Y. Kebukawa, T. Noguchi, D. Johnson, M.C. Price, K.H. Harriss, M.E. Zolensky, M.M. Grady, Organic matter and water from asteroid Itokawa. Sci. Rep. **11**, article 5125 (2021). https://doi.org/10.1038/s41598-021-84517-x

J.B. Cobb Jr., *Whitehead Wordbook: A Glossary with Alphabetical Index to Technical Terms in Process and Reality* (P&F Press, Claremont, CA, 2008)

K. Creath, G.E. Schwartz, Biophoton interaction in biological systems: evidence of photonic info-energy transfer? in *Proceedings SPIE*, vol. 5866. The Nature of Light: What Is a Photon? (2005), pp. 338–347. https://doi.org/10.1117/12.620846.

M.J. Crowe, *The Extraterrestrial Life Debate: Antiquity to 1915; A Source Book.* With a commentary by M.J. Crowe (University of Notre Dame Press, Notre Dame, IN, 2008)

Daodejing, 《道德經》 [*Tao Te Ching*] [= *Laozi*, 《老子》], Chinese Text Project (n.d.). https://ctext.org/dao-de-jing. Accessed 12 Sept 2022

M. Davis, *The Universal Computer: The Road from Leibniz to Turing*, 3rd edn. (CRC Press/Taylor & Francis Group, Boca Raton, 2018)

E.V. de Rougé, H. Brugsch, Stück eines ägyptischen Hymnus an die Sonne. Zeitschrift der Deutschen Morgenländischen Gesellschaft **4**, 374–377 (1850)

X. Delamarre, *Dictionnaire de la langue gauloise: une approche linguistique du vieux-celtique continental*, 2ème édn. (Erranc, Paris, 2003)

Deng Mu, 邓牧, 《伯牙琴》, 正文·补遗: 超然馆记, 國學大師 Guoxuedashi.net (n.d.). http://www.guoxuedashi.net/a/901haxl/20733z.html. Accessed 3 Jan 2022

Der Bundehesh. Herausgegeben, transkribiert, übersetzt und mit Glossar versehen von F. Justi (F.C.W. Vogel, Leipzig, 1868)

Diogenes Laertius, *Leben und Meinungen berühmter Philosophen*, Band 1: *Buch I-VII*. Übersetzt und erläutert von O. Apelt. Philosophische Bibliothek, Band 53 (F. Meiner, Leipzig, 1921)

Diogenes Laertius, *Lives of Eminent Philosophers*, vol. 2: *Books 6–10*. Greek-English edition, with an English translation by R.D. Hicks. The Loeb Classical Library, vol. 185 (Harvard University Press, Cambridge, MA; W. Heinemann, London, 1959)

M. Drieschner (ed.), *Carl Friedrich von Weizsäcker: Major Texts in Philosophy.* Springer Briefs on Pioneers in Science and Practice: Texts and Protocols, vol. 23 (Springer, Cham, 2014)

Epicurus, *The Extant Remains.* With short critical apparatus, translation, and notes edited by C. Bailey (Clarendon Press, Oxford, 1926)

A. Farhadi, C. Forsyth, A. Banan, M. Shaikh, P. Engen, J.Z. Fields, A. Keshavarzian, Evidence for non-chemical, non-electrical intercellular signaling in intestinal epithelial cells. Bioelectrochemistry **71**, 142–148 (2007). https://doi.org/10.1016/j.bioelechem.2007.03.001

J.G. Fichte, *The Science of Knowing: J. G. Fichte's 1804 Lectures on the Wissenschaftslehre.* Translated by W.E. Wright (State University of New York Press, Albany, NY, 2005)

J. Gebser, *Ursprung und Gegenwart*, Band 1: *Die Fundamente der aperspektivischen Welt* (Deutscher Taschenbuch Verlag, München, 1973)

R.L. Goerwitz III, Extraterrestrial life: a study of the intellectual context of Emanuel Swedenborg's earths in the universe. New Philos. **88**, 417–446 (1985)

A. Grandpierre, Cosmic life forms, in *From Fossils to Astrobiology: Records of Life on Earth and Search for Extraterrestrial Biosignatures*, ed. by J. Seckbach, M. Walsh (Springer, Dordrecht, 2008), pp. 369–385. https://doi.org/10.1007/978-1-4020-8837-7

A. Grandpierre, The fundamental biological activity of the universe, in *Eco-Phenomenology: Life, Human Life, Post-Human Life in the Harmony of the Cosmos*, ed. by W.S. Smith, J.S. Smith, D. Verducci. Analecta Husserliana, vol. 121 (Springer, Cham, 2018), pp. 115–140. https://doi.org/10.1007/978-3-319-77516-6

G. Günther, Life as poly-contexturality, in *Beiträge zur Grundlegung einer operationsfähigen Dialektik*, Band 2: *Wirklichkeit als Poly-Kontexturalität; Reflexion – Logische Paradoxie – Mehrwertige Logik – Denken – Wollen – Proemielle Relation – Kenogrammatik – Dialektik der natürlichen Zahl – Dialektischer Materialismus*, von G. Günther (F. Meiner, Hamburg, 1979), pp. 283–306

G. Günther, Das Problem einer transklassischen Logik, in *Beiträge zur Grundlegung einer operationsfähigen Dialektik*, Band 3: *Philosophie der Geschichte und der Technik; Wille – Schöpfung – Arbeit, Strukturanalyse der Vermittlung, Mehrwertigkeit – Stellen- und Stellenwertlogik, Kenogrammatik – Theorie der Zeit*, von G. Günther (F. Meiner, Hamburg, 1980), pp. 73–94

G. Günther, *Idee und Grundriss einer nicht-aristotelischen Logik: Die Idee und ihre philosophischen Voraussetzungen*. Mit einem Anhang "Das Phänomen der Orthogonalität" und mit einem Fragment aus dem Nachlass "Die Metamorphose der Zahl" (F. Meiner, Hamburg, 1991)

G. Günther, *Grundzüge einer neuen Theorie des Denkens in Hegels Logik*. Nach der Ausgabe der 2. Auflage (F. Meiner, Hamburg, 1978). Vordenker, Edition Seminartexte von J. Paul (2005). http://www.vordenker.de/ggphilosophy/theor-hegel-logik.pdf. Accessed 17 Jan 2022

G. Günther, *Der Tod des Idealismus und die letzte Mythologie*. Manuskript (Fragment), herausgegeben von R. Kaehr. Vordenker (n.d.). https://www.vordenker.de/ggphilosophy/tod-ideal.htm. Accessed 18 Sept 2022

P. Hadot, *The Veil of Isis: An Essay on the History of the Idea of Nature*. Translated by M. Chase (Belknap Press of Harvard University Press, Cambridge, MA, 2006)

G.W.F. Hegel, *Wissenschaft der Logik I; Erster Teil: Die objective Logik, Erstes Buch*. Werke, Band 5. Suhrkamp Taschenbuch Wissenschaft, Band 605 (Suhrkamp, Frankfurt am Main, 1986)

M. Heidegger, *Sein und Zeit*, 11. Auflage (Max Niemeyer, Tübingen, 1967)

D. Hertzer, *Das alte und das neue Yijing: Die Wandlungen des Buches der Wandlungen*. Diederichs Gelbe Reihe China, Band 126 (Diederichs, München, 1996).

M. Hess, *Dynamische Stofflehre; I. Kosmischer Theil: Allgemeine Bewegungserscheinungen und ewiger Kreislauf des kosmischen Lebens* (S.M. Hess, Paris, 1877)

D.T. Holden, N.M. Morato, R. Graham Cooks, Aqueous microdroplets enable abiotic synthesis and chain extension of unique peptide isomers from free amino acids. Proc. Natl. Acad. Sci. **119**(42), e2212642119 (2022). https://doi.org/10.1073/pnas.2212642119

B. Hoppe, Tellurisch; Tellurismus, in *Historisches Wörterbuch der Philosophie*, Band 10: *St-T*. Völlig neubearbeite Ausgabe des *Wörterbuchs der philosophischen Begriffe* von R. Eisler, herausgegeben von J. Ritter und K. Gründer (Wissenschaftliche Buchgesellschaft, Darmstadt, 1998), pp. 979–981

F. Hoyle, *The Intelligent Universe* (Holt, Rinehart and Winston, New York, 1983)

M. Isabel, Like supernovas, the human body emits bursts of light before dying. Cultura Colectiva: Inspiramos ideas, movemos mundos (2021). https://culturacolectiva.com/technology/human-body-emits-light-like-supernovas-before-dying/. Accessed 13 Oct 2022

K. Jaspers, *The Origin and Goal of History*, 3rd edn. Translated from the German by M. Bullock (Yale University Press, New Haven, 1965)

C.H. Kahn (ed.), *The Art and Thought of Heraclitus: An Edition of the Fragments with Translation and Commentary*, 9th edn. (Cambridge University Press, Cambridge, 2001)

I. Kant, *Allgemeine Naturgeschichte und Theorie des Himmels oder Versuch von der Verfassung und dem mechanischen Ursprunge des ganzen Weltgebäudes, nach Newtonischen Grundsätzen abgehandelt*. Das Bonner Kant Corpus (n.d. (a) [first 1755]), https://korpora.zim.uni-duisburg-essen.de/kant/aa01/256.html. Accessed 8 Jan 2022

I. Kant, *Kritik der praktischen Vernunft*. Das Bonner Kant Corpus (n.d. (b) [first 1788]), https://korpora.zim.uni-duisburg-essen.de/kant/aa05/161.html, https://korpora.zim.uni-duisburg-essen.de/kant/aa05/162.html. Accessed 8 Jan 2022

I. Kant, *Die Religion innerhalb der Grenzen der reinen Vernunft*. Das Bonner Kant Corpus (n.d. (c) [first 1793]), https://korpora.zim.uni-duisburg-essen.de/kant/aa06/100.html. Accessed 8 Jan 2022

I. Kant, *Metaphysik der Sitten*. Das Bonner Kant Corpus (n.d. (d) [first 1797]), https://korpora.zim.uni-duisburg-essen.de/kant/aa06/444.html. Accessed 8 Jan 2022

I. Kant, *Anthropologie in pragmatischer Hinsicht*. Das Bonner Kant Corpus (n.d. (e) [first 1798]), https://korpora.zim.uni-duisburg-essen.de/kant/aa07/331.html. Accessed 8 Jan 2022

I. Kant, *Logik*. Das Bonner Kant Corpus (n.d. (f) [first 1800]), https://korpora.zim.uni-duisburg-essen.de/kant/aa09/025.html. Accessed 8 Jan 2022

I. Kant, *Opus postumum*. Das Bonner Kant Corpus (n.d. (g)). https://korpora.zim.uni-duisburg-essen.de/kant/aa21/Inhalt21.html. Accessed on January 8, 2022

M. Kobayashi, D. Kikuchi, H. Okamura, Imaging of ultraweak spontaneous photon emission from human body displaying diurnal rhythm. PLoS ONE **4**(7), e6256 (2009). https://doi.org/10.1371/journal.pone.0006256

J. Kreuzer, *Gestalten mittelalterlicher Philosophie: Augustinus, Eriugena, Eckhart, Tauler, Nikolaus v. Kues* (W. Fink, München, 2000)

S.L. Kwee, *Methods of Comparative Philosophy* (printed at Offsetdrukkerij Dorsman, Scheveningen, 1953)

L. Landen, Biotic community—real or unreal: a philosophical dilemma. Worldviews **7**(1/2), 58–79 (2003). https://doi.org/10.1163/156853503321916219

W. Li, *Die christliche China-Mission im 17. Jahrhundert: Verständnis, Unverständnis, Missverständnis; Eine geistesgeschichtliche Studie zum Christentum, Buddhismus und Konfuzianismus*. Studia Leibnitiana Supplementa, Band 32 (Steiner, Stuttgart, 2000)

Liezi, 《列子》 [*The Book of Master Lie*], Chinese Text Project (n.d.). https://ctext.org/liezi. Accessed 12 Sept 2022

Liji, 《禮記》 [*Book of Rites*], Chinese Text Project (n.d.). https://ctext.org/liji. Accessed 12 Sept 2022

M. Lingam, A. Loeb, *Life in the Cosmos: From Biosignatures to Technosignatures* (Harvard University Press, Cambridge, MA, 2021)

F. Lötzsch, Monade, Monas (griech. μονάς), in *Historisches Wörterbuch der Philosophie,* Band 6: *Mo-O*. Völlig neubearbeite Ausgabe des *Wörterbuchs der philosophischen Begriffe* von R. Eisler, herausgegeben von J. Ritter und K. Gründer (Wissenschaftliche Buchgesellschaft, Darmstadt, 1984), pp. 114–117

N. Luhmann, *Ecological Communication*. Translated by J. Bednarz Jr. (University of Chicago Press, Chicago; Polity Press, Cambridge, 1989)

N. Luhmann, Die Paradoxie der Form, in *Aufsätze und Reden*, von N. Luhmann, herausgegeben von O. Jahraus. Reclams Universal-Bibliothek, Band 18141 (Philipp Reclam, Stuttgart, 2001), pp. 243–261

N. Luhmann, *Ideenevolution: Beiträge zur Wissenssoziologie.* Herausgegeben von A. Kieserling. Suhrkamp Taschenbuch Wissenschaft 1870 (Suhrkamp, Frankfurt am Main, 2008)

H.R. Maturana, The biological foundations of self consciousness and the physical domain of existence, in *Beobachter: Konvergenz der Erkenntnistheorien?*, von N. Luhmann, H.R. Maturana, M. Namiki, V. Redder und F. Varela, 3. Auflage (W. Fink, Paderborn, 2003), pp. 47–117

S. Mayburov, Photonic communications and information encoding in biological systems (2012) arXiv:1205.4134v1 [q-bio.OT]. https://doi.org/10.48550/arXiv.1205.4134

K. McCormick, The spooky quantum phenomenon you've never heard of. Quanta Magazine (2022). https://www.quantamagazine.org/the-spooky-quantum-phenomenon-youve-never-heard-of-20220622/. Accessed 17 2022

B. Mojsisch, Notiz ›*docta ignorantia*‹. Bochumer Phil. Jrbh. **6**(1), 240–241 (2001). https://doi.org/10.1075/bpjam.6.11moj

C. Moro, A. Liebert, C. Hamilton, N. Pasqual, G. Jeffery, J. Stone, J. Mitrofanis, The code of light: do neurons generate light to communicate and repair? Neural Regen. Res. **17**(6), 1251–1252 (2022). https://doi.org/10.4103/1673-5374.327332

M. Müller, Preface to the Sacred Books of the East (1879a). In *Müller* (1879b), pp. ix–ci

M. Müller (ed.), *The Upanishads, Part I: The Khândogya-Upanishad, the Talavakâra-Upanishad, the Aitareya-Âranyaka, the Kaushîtaki-Brâhmana-Upanishad and the Vâgasaneyi-Samhitâ-Upanishad*. Translated by M. Müller. Sacred Books of the East, vol. 1 (Clarendon, Oxford, 1879b)

J. Needham, *Science and Civilisation in China*, vol. 2: *History of Scientific Thought*, 5th edn. (Cambridge University Press, Cambridge, 1977)

R.C. Neville, Philosophische Grundlagen und Methoden der Komparativen Theologie, in *Komparative Theologie: Interreligiöse Vergleiche als Weg der Religionstheologie*, herausgegeben von R. Bernhardt und K. v. Stosch. Beiträge zu einer Theologie der Religionen, Band 7 (TVZ, Zürich, 2009), pp. 35–54

Nicholas of Cusa [= Nicolaus Cusanus], *Complete Philosophical and Theological Treatises of Nicholas of Cusa,* vol. 2. Translated by J. Hopkins (A.J. Banning, Minneapolis, MN, 2001), electronic version: Cusanus-Portal. http://www.cusanus-portal.de/content/fw.php?werk=13&fw=170&ln=hopkins. Accessed 10 Jan 2022

Nicolaus de Cusa [= Nicolaus Cusanus], *Ὀφθαλμός Ἁπλόος or The single Eye: Entituled the Vision of God wherein is infolded the mistery of divine presence, so to be in one place finitely in appearance, as yet in every place no lesse present, and whilst Hee is here, Hee is universally every where infinitely himself.* Edited by G. Randall (J. Streater, London, 1646)

Nicolaus de Cusa [= Nicolaus Cusanus], *De docta ignorantia*, in *Nicolai de Cusa opera omnia iussu et auctoritate academiae litterarum Heidelbergensis ad codicum fidem edita*, vol. I, ediderunt E. Hoffmann et R. Klibansky (F. Meiner, Lipsiae, 1932)

Nicolaus de Cusa [= Nicolaus Cusanus], *De filiatione Dei*, in *Nicolai de Cusa opera omnia iussu et auctoritate academiae litterarum Heidelbergensis ad codicum fidem edita*, vol. IV: *Opuscula I. De deo abscondito, De quaerendo Deum, De filiatione Dei, De dato patris luminum, Coniectura de ultimis diebus, De genesi*, edidit P. Wilpert (F. Meiner, Hamburgi, 1959), pp. 39–64

Nicolaus de Cusa [= Nicolaus Cusanus], *De coniecturis*, in *Nicolai de Cusa opera omnia iussu et auctoritate academiae litterarum Heidelbergensis ad codicum fidem edita*, vol. III, ediderunt J. Koch et K. Bormann, J.G. Senger comite (F. Meiner, Hamburgi, 1972)

Nicolaus de Cusa [= Nicolaus Cusanus], *De beryllo*, in *Nicolai de Cusa opera omnia iussu et auctoritate academiae litterarum Heidelbergensis ad codicum fidem edita*, vol. XI, 1, editionem funditus renovatam atque instauratam curaverunt J.G. Senger et K. Bormann (F. Meiner, Hamburgi, 1988)

Nicolaus de Cusa [= Nicolaus Cusanus], *De visione Dei*, in *Nicolai de Cusa opera omnia iussu et auctoritate academiae litterarum Heidelbergensis ad codicum fidem edita*, vol. VI, edidit H.D. Riemann (F. Meiner, Hamburgi, 2000)

B. Nielsen, *A Companion to Yi Jing Numerology and Cosmology: Chinese Studies of Images and Numbers from Han (202 BCE-220 BCE) to Song (960–1279 CE)* (Routledge Curzon, London, 2003)

L. Oken, *Lehrbuch der Naturphilosophie* (Frommann, Jena, 1809–11)

J. Opsomer, C. Steel, Introduction, in *Proclus* (2003), pp. 1–53

M. Planck, *Scientific Autobiography and Other Papers*. With a memorial address on M. Planck by M. von Laue, translated from German by F. Gaynor (Williams & Norgate, London, 1950)

Plato, Apology, in *The Dialogues of Plato in Five Volumes*, vol. 2. Translated into English with analyses and introduction by B. Jowett (Clarendon, Oxford, 1892), pp. 95–135

Plato, *Euthyphro – Apology – Crito – Phaedo – Phaedrus*. Greek-English edition, with an English translation by H.N. Fowler. Plato, vol. 1 (W. Heinemann, London; MacMillan, New York, 1913)

Plato, *Timaeus and Critias*. Translated by R. Waterfield, with an introduction and notes by A. Gregory. Oxford World's Classics (Oxford University Press, Oxford, 2008)

Plotinus, *The Enneads*. Edited by L.P. Gerson, translated by G. Boys-Stones, J.M. Dillon, L.P. Gerson, R.A.H. King, A. Smith, J. Wilberding (Cambridge University Press, Cambridge, 2018)

F.-A. Popp, About the coherence of biophotons. Stealthskater (n.d. [first 1998]), http://stealthskater.com/Documents/Consciousness_31.pdf. Accessed 19 Jan 2022

Proclus, *On the Existence of Evils*. Translated by J. Opsomer, C. Steel. Ancient Commentators on Aristotle (Cornell University Press, Ithaca, NY, 2003)

W. Rasch, Theories of complexity, complexities of theory: Habermas, Luhmann, and the study of social systems. Ger. Stud. Rev. **14**(1), 65–83 (1991). https://doi.org/10.2307/1430154

Refutation of All Heresies. Translated with an introduction and notes by M.D. Litva. Writings from the Graeco-Roman World, vol. 40 (SBL Press, Atlanta, 2016)

H. Romijn, Are virtual photons the elementary carriers of consciousness? J. Conscious. Stud. **9**(1), 61–81 (2002)

W. Ruben, *Die Philosophen der Upanishaden* (A. Francke, Bern, 1947)

C.L. Sanders, Speculations about bystander and biophotons. Dose Response **12**(4), 515–517 (2014). https://doi.org/10.2203/dose-response.14-002.Sanders

B.-A. Scharfstein, *A Comparative History of World Philosophy: From the Upanishads to Kant* (State University of New York Press, Albany, NY, 1998)

E. Schrödinger, *What is Life? The Physical Aspect of the Living Cell; With Mind and Matter and Autobiographical Sketches*, 14th edn. (Cambridge University Press, Cambridge, 2013)

M. Schönberger, *The I Ching & The Genetic Code: The Hidden Key to Life*. Foreword by Lama Anagarika Govinda, translated [from the German] by D.Q. Stephenson (Aurora Press, Santa Fe, 1992)

Shizi, 尸子, 中国哲学命题大全 [Comprehensive Compilation of Chinese Philosophical Propositions] (n.d.). http://chinese.zhexue.org/philosophy/18177. Accessed 18 Jan 2022

J. Silk, *The Infinite Cosmos: Questions from the Frontiers of Cosmology* (Oxford University Press, Oxford, 2006)

Simplicius, *On Aristotle On the Heavens 1.5–9*. Translated by R.J. Hankinson. Ancient Commentators on Aristotle (Bloomsbury, London, 2014)

The emerging field of biophotonic communication, MIT Technology Review (2009), https://www.technologyreview.com/2009/11/19/123329/the-emerging-field-of-biophotonic-communication/. Accessed 13 Oct 2022

E. Theodossiou, P. Mantarakis, M.S. Dimitrijevic, V.N. Manimanis, E. Danezis, From the infinity (apeiron) of Anaximander in ancient Greece to the theory of infinite universes in modern cosmology. Astron. Astrophys. Trans. **27**(1), 162–176 (2010/2011)

B. Waldenfels, *In den Netzen der Lebenswelt* (Suhrkamp, Frankfurt am Main, 1985)

T. Walker Jr., N.C. Wickramasinghe, *The Big Bang and God: An Astrotheology* (Palgrave MacMillan, Houndsmills, Basingstoke, 2015)

Wang Shouren, 王守仁 [= Wang Yangming, 王陽明], 《傳習錄》 (中) [*Records of Teaching and Learning*, part 2 of 3], 《王文成公全書》, 第2冊, 王雲五主編 (商務印書館, 上海, 1933a), pp. 37–82

Wang Shouren 王守仁 [= Wang Yangming, 王陽明], 《傳習錄》 (下) [*Records of Teaching and Learning*, part 3 of 3], 《王文成公全書》, 第3冊, 王雲五主編 (商務印書館, 上海, 1933b), pp. 1–48

Z. Wang, N. Wang, Z. Li, F. Xiao, J. Dai, Human high intelligence is involved in spectral redshift of biophotonic activities in the brain. Proc. Natl. Acad. Sci. **113**(31), 8753–8758 (2016). https://doi.org/10.1073/pnas.1604855113

C.F. von Weizsäcker, *Die Geschichte der Natur: Zwölf Vorlesungen* (S. Hirzel, Stuttgart, 1948)

C.F. von Weizsäcker, *Die Einheit der Natur* (C. Hanser, München, 1971)

A.N. Whitehead, *Process and Reality: An Essay in Cosmology; Gifford Lectures Delivered in the University of Edinburgh During the Session 1927–28*. Corrected edition by D.R. Griffin, D.W. Sherburne (The Free Press/Macmillan Publishing, New York, 1978)

N.C. Wickramasinghe, Comets, panspermia, culture, and prejudice. Asia Pac. Biotech News **25**(10), 26–31 (2021). https://doi.org/10.1142/S0219030321000100

L. Wittgenstein, *Philosophische Untersuchungen. Philosophical Investigations*. The German text, with an English translation by G.E.M. Anscombe, P.M.S. Hacker, J. Schulte, revised 4th ed. by P.M.S. Hacker, J. Schulte (Wiley-Blackwell, Chichester, West Sussex, 2009)

M. Woermann, *Bridging Complexity and Post-Structuralism: Insights and Implications* (Springer, Cham, 2016)

G. Wohlfart, *Der Philosophische Daoismus: Philosophische Untersuchungen zu Grundbegriffen und komparative Studien unter besonderer Berücksichtigung des Laozi (Lao-tse)*. Reihe für asiatische und komparative Philosophie, Band 5 (Edition Chora, Köln, 2001)

W. Wollaston, *The Religion of Nature Delineated*, 7th ed. (R. Urie, Glasgow, 1746)

J. Wolter, *Apparitio Dei: Der theophanische Charakter der Schöpfung nach Nikolaus von Kues*. Beiträge zur Geschichte der Philosophie und Theologie des Mittelalters: Neue Folge, Band 67 (Aschendorff, Münster, 2004)

T. Wright, *An Original Theory or New Hypothesis of the Universe Founded Upon the Laws of Nature and Solving by Mathematical Principles the General Phaenomena of the Visible Creation; and particularly The Via Lactea* (H. Chapelle, London, 1750)

D. Wujastyk, On attaining special powers through *Rasāyana* therapies in Sanskrit medical literature, in *Body and Cosmos: Studies in Early Indian Medical and Astral Sciences in Honor of Kenneth G. Zysk*, ed. by T.L. Knudsen, J. Schmidt-Madsen, S. Speyer (Brill, Leiden, 2021), pp. 140–165

Yijing, 《周易》 [*Book of Changes*], Chinese Text Project (n.d.). https://ctext.org/book-of-changes/shuo-gua. Accessed 21 Jan 2022

E.V. Zenker, *Histoire de la philosophie chinoise*. Traduit de l'allemand par G. Lepage et Y. Le Lay (Editions Payot, Paris, 1932)

Zhang Heng, 张衡, 中国哲学命题大全 [Comprehensive Compilation of Chinese Philosophical Propositions] (n.d.). http://chinese.zhexue.org/philosophy/18193. Accessed 18 Jan 2022

W. Zhang, H. Shen, The philosophy of the *Yijing* and Jungian analytical psychology: a comparative study. J. Analyt. Psychol. **66**(4), 969–990 (2021). https://doi.org/10.1111/1468-5922.12714

David Bartosch (Germany) is an internationally recognized expert in Chinese, German, and other philosophical traditions. He specializes in transcultural comparative philosophy and has applied this knowledge to the fields of intercultural relations and ecological thought. He is a distinguished research fellow at Beijing Normal University at Zhuhai, China.

The manufacturer's authorised representative in the EU is Springer Nature Customer Service Centre GmbH, Europaplatz 3, 69115 Heidelberg, Germany. If you have any concerns regarding our products, please contact ProductSafety@springernature.com

Printed and bound by CPI Group (UK) Ltd, Croydon, CR0 4YY
30/09/2026
02225318-0001